Handbook of
Environmental Acoustics

Handbook of
Environmental Acoustics

James P. Cowan

VNR VAN NOSTRAND REINHOLD
New York

Library of Congress Catalog Card Number 93-35640
ISBN 0-442-01644-1

I(T)P   Van Nostrand Reinhold is an International Thomson Publishing company.
ITP logo is a trademark under license.

Printed in the United States of America

Van Nostrand Reinhold   ITP Germany
115 Fifth Avenue   Königswinterer Str. 418
New York, NY 10003   53227 Bonn
                                 Germany

International Thomson Publishing International Thomson Publishing Asia
Berkshire House,168-173   38 Kim Tian Rd., #0105
High Holborn, London WC1V 7AA   Kim Tian Plaza
England   Singapore 0316

Thomas Nelson Australia   International Thomson Publishing Japan
102 Dodds Street   Kyowa Building, 3F
South Melbourne 3205   2-2-1 Hirakawacho
Victoria, Australia   Chiyada-Ku, Tokyo 102
                                 Japan

Nelson Canada
1120 Birchmount Road
Scarborough, Ontario
M1K 5G4, Canada

16 15 14 13 12 11 10 9 8 7 6 5 4 3 2 1

Library of Congress Cataloging in Publication Data
Cowan, James P.
    Handbook of environmental acoustics / James P. Cowan.
        p.   cm.
    Includes bibliographical references and index.
    ISBN 0-442-01644-1
    1. Architectural acoustics. I. Title.
NA2800.C64 1994
729'.29—dc20
                                        93-35640
                                        CIP

To Ida Levy, Alfred, Lorraine, and Lynn Cowan,
without whose loving support
this book would not have been possible

Contents

Preface

In the early 1980s, the U.S. Environmental Protection Agency Office of Noise Abatement and Control (ONAC) estimated that 87% of the population in urban areas were exposed to noise levels above those that are recommended to protect the population from adverse health and safety effects. Since that time, urban areas have expanded their borders to encompass former rural areas, and common noise sources, although quieted slightly by technology, have multiplied and have been located closer to many residential and sensitive areas. The consensus is that community noise levels today are at least as high as they were in the early 1980s, and are probably higher.

With the expansion of transportation system routes and industrial operations, noise problems are not restricted to urban areas. Human activities, both professional and recreational, may also generate noise levels that can intrude on other people. The bottom line is that no area is immune from noise-related problems.

Noise seems to be an environmental stressor that is low on most priority lists because it shows no obvious immediate health risks. However, in addition to the known fact that noise can cause hearing loss, noise has been linked with many stress-related illnesses. As environmental consciousness becomes more prevalent, noise issues will follow.

Every discipline has its own esoteric language that sets its experts apart from the rest of the population, and noise is no exception to this rule. In addition to its own language, noise assessment also has a different mathematical way of describing its parameters. The intent of this book is to break through these language barriers and provide the most relevant information that would be needed to understand noise issues from a practical viewpoint.

There are several texts available that deal with noise on a level appropriate for an engineer or physicist concerned with derivations or complex analysis. This book is meant for everyone else: it provides a basic understanding of noise issues and terminology without high-level theory or mathematics. Although I obtained an advanced degree in acoustics, I feel my practical knowledge in the field was acquired through real-life encounters following my graduation. While teaching classes in this area, I found no practical text that explained the most important concepts to students new to the field. This book provides the kind of information I wanted to have available when I was first learning about acoustics.

All relevant topics dealing with noise issues are provided, along with summaries of regulations, guidelines, and standards commonly used for noise assessment. A glossary is provided at the end of the text, along with a model noise ordinance that includes all commonly accepted principles. The first five chapters have review questions to reinforce the most important concepts discussed in those chapters. There are no trick questions or questions that would involve looking elsewhere for solution paths. They are simply questions that will help you to review the principles discussed.

I would like to acknowledge the fine work of Karin Greene and Stacy Alabardo in preparing most of the artwork for this book. Sincere thanks is extended to Ike Richman, the Spectrum, Spectrum president Carl Hirsch, and Ron Howard for graciously allowing me to sample noise levels at the spectator events referenced in this book. The support of manufacturers and vendors who provided photographs and charts is also greatly appreciated.

Handbook of
Environmental Acoustics

1

Acoustic Terminology

INTRODUCTION

Noise, in its simplest definition, is unwanted sound. The adage of one man's noise being another's music is the basis for environmental noise concerns. Although high levels of noise may cause hearing loss, the levels usually associated with environmental noise assessments are below the hazardous range. They should not be overlooked for this reason, however, because they can cause stress-related illnesses, sleep deprivation, communication interference, and may interrupt activities requiring concentration. Such problems are often viewed as by-products of our technologically advancing and expanding society.

Because noise usually does not cause immediate permanent damage and is not life-threatening in the way many other environmental hazards can be, it is usually placed low on priority lists until an emergency situation arises. Environmental noise became an issue with the opening of the Office of Noise Abatement and Control (ONAC) by the federal Environmental Protection Agency (EPA) in the early 1970s. However, 10 years later funding was dropped for the ONAC and, with it, most concerns for noise control. Now that 10 years have passed since the closing of the ONAC, there is talk of reopening it. Whether the ONAC reopens or not, noise is an issue that must be contended with as long as it exists and people can hear. Among those dealing with noise issues, there is much confusion over terminology and control methods. Because local certification is unavailable in this discipline, self-proclaimed experts may worsen noise environments through their recommendations.

This book has been written as a reference guide for the reader who needs a basic understanding of the discipline but has little or no background in

1

acoustics. The book is also a handy reference for the reader to use to ensure that the hired "expert" truly deserves the title of acoustic or noise control engineer. The chapters are organized in terms of the major areas of concern, the basics of which must be understood when dealing with an environmental noise issue. Questions are included at the end of each chapter to summarize the topics covered and to ensure full comprehension of those topics. Charts and tables are also included for checklists and calculations.

All references used in this book are to current versions of the most relevant publications from the agencies, societies, and organizations responsible for research and criterion establishment in environmental noise. These provide a consensus agreement from the leading experts in their respective areas of expertise and do not reflect the opinions of isolated individuals. For this reason, the articles written by individuals that are referenced in this book are included only because they are endorsed as having the information that is most widely accepted by public and private agencies. The author therefore believes that the information contained in this book provides the reader with the most current, unbiased information available in this area.

ACOUSTIC PARAMETERS

Acoustics is the science of sound. Noise, being undesirable sound, is therefore a subdiscipline of acoustics and is described in terms of acoustic parameters. Two key parameters must be understood when dealing with any acoustic concern. These are frequency and wavelength, quantities that describe the nature of pressure fluctuations in a medium (air, in this case) that are eventually interpreted as sound in the brain. Frequency and wavelength are related to each other through the speed of sound, which dictates the direction of sound travel (dictated by the variation in the speed of sound) and the time that sound arrives at a listener's ears.

Frequency

Visualize a piston, set horizontally inside a closed cylinder, moving side to side at a constant rate of n times each second. The volume of air inside the cylinder is being compressed when the piston is extended to its farthest position in one direction and given room to expand when the piston is at its farthest position in the other direction. The back-and-forth motion of the piston in the cylinder therefore sets up pressure variations, or waves, in the cylinder that travel, or propagate, along its length. One complete cycle of pressure variation would correspond to one back-and-forth motion of the piston. The rate at which this occurs, in cycles of pressure variation per second, is known as the frequency of the wave. The units of cycles per second

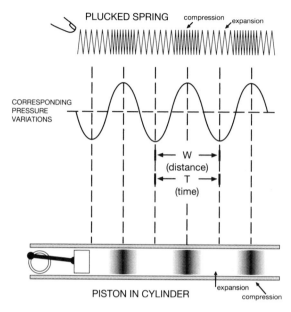

FIGURE 1-1.   Illustrations of wave motion analogous to sound wave motion for a single frequency or pure tone. *w*, Wavelength; *T*, period (reciprocal of the frequency).

are usually denoted as hertz (abbreviated Hz), named for the German physicist Heinrich Rudolph Hertz (1857–1894).

The pressure inside the cylinder, set up by the motion of the piston, would change with distance from the piston in a sinusoidal (like a sine wave) pattern. In this case, the pressure pattern, when plotted versus distance along the cylinder from the piston, would repeat itself after each complete cycle. The distance between repeating sections of the pressure wave is known as the wavelength of the wave. A diagram of this situation is shown in Fig. 1-1 to clarify this concept. Figure 1-1 also shows another illustration of this wave motion, such as would occur when a spring is plucked. The areas of compression and expansion along the spring clearly show the pressure wave traveling along the spring. The pressure variations described above, when occurring at levels and frequencies to which our hearing mechanisms are sensitive, are interpreted as sound in our brains.

If the sound source is a point source, that is, the size of the sound source is small compared with the distance between the source and the listener, and there are no surfaces near it to obstruct its travel, the pressure wave would radiate spherically in a sinusoidal pattern away from the source. A way to visualize this acoustic pressure wave pattern is by looking at a calm body

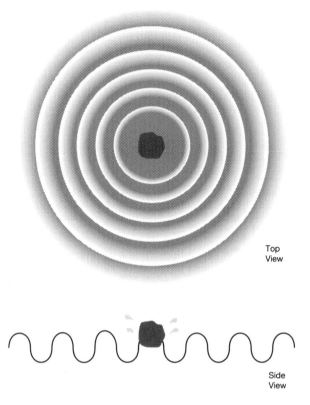

FIGURE 1-2.   An illustration of spherical wave motion caused by a pebble falling into a still body of water.

of water just after a pebble has been dropped into it. If the observer were at the top level of the water (where it meets the air), looking across the water between the pebble impact location and one farther away, he would see the peak and valley pattern in the water radiating in circular patterns away from the initial impact location. This is illustrated in Fig. 1-2.

A sound composed of a single frequency, as described in the piston example above, is known as a pure tone. Pure tones rarely exist in nature except under ideal conditions, such as in musical instruments and electronically synthesized sounds. When pure tones exist in environmental sound, they tend to be more annoying to people than other typical sounds. Typical sounds that we hear are composed of many frequencies at different levels. We hear the combination as a single sound interpreted by the brain.

Human Hearing Range

It is generally accepted that humans can hear frequencies between 20 and 20,000 Hz; however, we are most sensitive to sounds with frequencies between 500 and 4000 Hz. As frequencies fall below 500 Hz and rise above 4000 Hz, our hearing mechanism reduces, or attenuates, the sound level that we hear. This means that a certain level of sound would be perceived, when its dominant frequencies shift out of the 500- to 4000-Hz range, as quieter than it actually is. Our hearing mechanisms amplify sounds around 2000 Hz because the open-ended tube formed by the ear canal and terminated by the eardrum acts as a tuned organ pipe. Just as the frequency, perceived as pitch, of sound emitted from an organ pipe depends on the length and diameter of the pipe, our ear canals are tuned to frequencies around 2000 Hz by nature of their size and shape.

In musical terms, middle C on the piano scale corresponds to roughly 250 Hz and an octave change corresponds to a doubling or halving of frequency. The dominant energy components of human speech are in the 500- to 2000-Hz range, with vowel sounds in the low end and consonants in the high end of that range.

Infrasound

Sound having dominant frequency components below 20 Hz is known as infrasound. Infrasound, at very high levels, can be the most dangerous sound to which humans could be exposed. The danger is caused by the phenomenon of resonance. Every physical thing has a resonance frequency associated with it that depends on its material composition. When the object is exposed to high levels of sound dominant in its resonance frequency range, it tends to oscillate, or vibrate, at a frequency similar to the exciting tone and at a larger amplitude than it would if the exciting tone were out of the resonance frequency range. Our internal organs resonate at infrasonic frequencies between 5 and 15 Hz. This is why, when the sound level is high enough, sounds dominant in this frequency range cannot be heard but can be felt as vibrations in our bodies. This vibration of internal organs may damage these tissues and cause circulatory and stress-related illnesses. A clear link between infrasound exposure and human health has not been established to date; however, studies on these effects are ongoing and it is advisable to avoid unnecessary exposure to any source that causes vibrations that can be felt in any part of the body.

The wavelength of a 20-Hz signal is more than 50 ft, as is shown in Fig. 1-3. Also, wavelength increases as frequency decreases. The large wavelengths associated with infrasound travel over longer distances and are more difficult to attenuate than higher frequencies. This adds to the potentiality of infrasonic hazards.

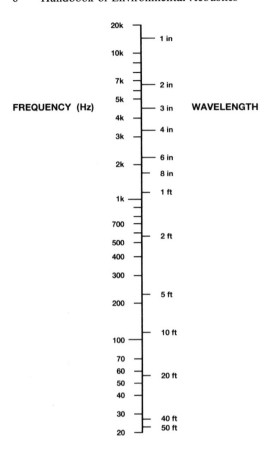

FREQUENCY (Hz)   WAVELENGTH

FIGURE 1-3. Correlation between frequency and wavelength at normal room temperature (70°F).

Ultrasound

On the other end of the sound frequency spectrum is ultrasound, having dominant frequency components above 20,000 Hz. Typical notation uses the symbol k (standing for kilo or 1000) to replace thousands of Hz by kHz (e.g., 20,000 Hz is denoted as 20 kHz). A 20-kHz wavelength is less than an inch with higher frequencies having even smaller wavelengths, as is shown in Fig. 1-3. Because many ultrasonic wavelengths can fit into a small area, high-level ultrasonic beams can be formed that have been known to do anything from cleaning teeth to drilling sidewalks. Ultrasound attenuates rapidly with distance from the source and therefore is usually not dangerous unless a person is in contact with a high-level beam source.

Low levels of ultrasound are used extensively in the medical profession to view internal organs, blood flow, and fetuses without the potential hazards involved in using X rays. The principle used in image generation is to gather reflected beams of focused sound off of materials of different densities. Because different materials change sound speed and direction when sound waves travel through them, the ultrasonic devices emit a sound and sense its characteristics after being reflected back to the instrument from the tissues in question. This reflected sound, when compared with the emitted sound, is processed and transferred to a screen, where a picture of the materials is generated.

Focused beams of ultrasound are also used successfully by the medical profession for such procedures as breaking up kidney stones and cleaning plaque from clogged arteries.

Wavelength

The wavelength is the distance between successive repeating parts of the pure tone (single frequency), sinusoidal acoustic pressure wave. This is shown in Fig. 1-1. The concept of wavelength becomes important when dealing with noise control because materials react differently to sounds having different wavelengths. This will be discussed in more detail later in this chapter and in Chapter 4.

Speed of Sound

The speed of sound is, as its name implies, the speed at which the acoustic pressure wave travels away from the sound source. It can be visualized as the speed of one of the waves traveling away from the point of impact of the pebble in the water analogy mentioned above. In the plucked spring analogy used in Fig. 1-1, the pressure wave can easily be seen traveling along the spring.

The mathematical relationship between frequency and wavelength is through the speed of sound. The basic equation of this relationship is:

$$c = w \times f \tag{1-1}$$

where c is the speed of sound, w is the wavelength, and f is the frequency. As long as the speed of sound is constant, this equation shows that wavelength and frequency have an inverse relationship—as frequency decreases, wavelength increases and as frequency increases, wavelength decreases—as was implied in the discussion above on infrasound and ultrasound. Figure 1-3 shows wavelengths with associated frequencies over the human hearing range. As can be seen from Fig. 1-3, we hear a large range of

TABLE 1-1   Speed of Sound in Common
Media at 70°F

Medium	c (ft/s)
Air	1,128
Sea water	4,920
Plexiglas	5,900
Concrete	11,200
Copper	11,500
Wood, marble	12,500
Steel	16,600
Aluminium	16,900
Glass	17,100
Gypsum	22,300

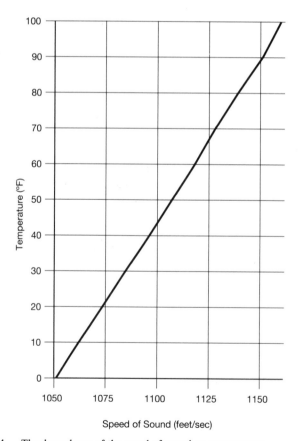

FIGURE 1-4.   The dependence of the speed of sound on temperature.

wavelengths, ranging from less than an inch at 20 kHz to more than 50 ft at 20 Hz.

The speed of sound is dependent on the medium the sound is traveling through and its temperature. The speed of sound is listed in Table 1-1 for common media at 70°F.

Figure 1-4 shows the variation of the speed of sound (in air) with temperature. For those interested in calculating the speed of sound at any temperature, Eq. (1-2) can be used:

$$c = 49.03 \times \sqrt{459.7 + °F} \quad \text{in ft/s} \tag{1-2}$$

SOUND PROPAGATION

Sound usually does not travel by line of sight, especially over distances of more than 200 ft. The medium of propagation is assumed to be air for this discussion. There are four principal methods by which sound changes direction of propagation as it travels through air. These phenomena—reflection, refraction, diffraction, and diffusion—occur when a sound wave encounters a change in medium, temperature, humidity, or wind currents.

The properties of sound travel are similar to those of light. Although sound and light are based on different types of energy, both have characteristics of frequency and wavelength. A principal difference lies in the different ranges of parameters to which optical and auditory organs are sensitive. Light is typically described in terms of its wavelength, whereas sound is typically described in terms of its frequency. For reference, the smallest wavelength that we can hear is on the order of 20,000 times larger than the largest wavelength that we can see.

Reflection

Reflection of sound occurs when sound waves bounce off a surface at the same angle, with respect to a line perpendicular to the surface, at which the sound was incident on the surface. In other words, the angle of incidence equals the angle of reflection, similar to the behavior of light when it encounters a mirror. Figure 1-5 illustrates this phenomenon. For clear reflection to take place, the reflecting surface must be larger than the wavelength of the sound wave incident on the surface.

Refraction

Refraction of sound is the bending of sound waves caused by propagation through continually varying media or medium conditions. In physics, acous-

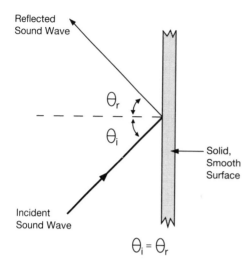

$$\theta_i = \theta_r$$

FIGURE 1-5.   Reflection of sound waves.

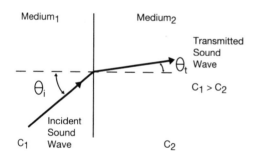

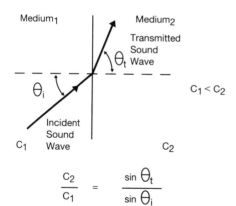

$$\frac{c_2}{c_1} = \frac{\sin \theta_t}{\sin \theta_i}$$

FIGURE 1-6.   Refraction of sound waves.

tic refraction is described by Snell's law as follows:

$$c_2/c_1 = \sin \theta_t/\sin \theta_i \tag{1-3}$$

where c_2 is the speed of sound in the transmitting medium, c_1 is the speed of sound in the incident medium, θ_t is the angle of transmission (with respect to the line perpendicular to the medium interface), and θ_i is the angle of incidence (with respect to the line perpendicular to the medium interface). Figure 1-6 illustrates this phenomenon.

Snell's law states that sound will change direction when traveling into a medium that conducts sound at a different speed. This can occur when a sound wave travels into a different medium or when a sound wave travels into a different condition in the same medium. For instance, as sound waves travel through the atmosphere, temperature, humidity, and wind current conditions change and, because the speed of sound in air is dependent on these factors, sound speed and direction change accordingly. As light bends when traveling through a prism, sound bends when traveling through the acoustic prism of varying atmospheric properties. Sound propagation in large outdoor areas will be discussed in more detail later in this chapter.

Diffraction

Diffraction is the phenomenon of sound waves apparently bending around partial barrier walls. This phenomenon occurs when the wavelength of sound incident on the barrier is comparable to or larger than the height and width dimensions of the barrier. It is diffraction, as is shown in Fig. 1-7, that gives highway noise barriers and open office plan cubicle barriers practical limits of attenuation. The specifics of these limits are discussed in Chapter 4. Visual privacy from a noise source shielded by a partial (open to the air on top) barrier does not imply acoustic privacy, because of the diffraction phenomenon.

Diffusion

Diffusion is the even spreading of sound that occurs when sound waves reflect off of convexly curved or uneven surfaces, as is shown in Fig. 1-8. The optical analogy to this would be the effect of a frosted light bulb. This effect is desirable in concert halls and recording studios to eliminate sharp echoes without eliminating the sound by absorbing it. By using diffusive surfaces, the sound is distributed evenly around the room, thus blending musical sounds over a broad area.

Diffusion is also used in recording studios to create the aural illusion of a space that is larger than it actually is. In this way, concert hall sound can

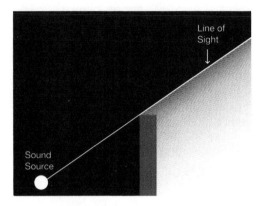

FIGURE 1-7.   Diffraction of sound waves around a barrier. The relative sound level is indicated by the degree of shading.

be generated in smaller rooms. A product line has been developed by P. D'Antonio (RPG Diffusor Systems, Inc., Upper Marlboro, MD) that capitalizes on this phenomenon. RPG Diffusor Systems, Inc., offers a wide variety of products based on wells of varying depths determined by established mathematical and scientific principles to generate an evenly diffuse field within a room. Sound waves incident on panels having these types of designs are diffused evenly throughout the room they are bordering. RPG stands for reflection phase grating, and Figs. 1-9 and 1-10 show sample panels and installations of these products. The general concept behind these diffusors

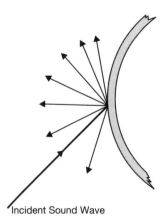

FIGURE 1-8.   Diffusion of sound waves incident on a convex surface.

FIGURE 1-9.   A diffusor panel. Note the uneven depth of wells in the cross-section. (Courtesy of RPG Diffusor Systems, Inc., 651-C Commerce Drive, Upper Marlboro, MD 20772. With permission.)

was originally derived by the German physicist Manfred Schroeder. His invention of the quadratic residue diffusor (denoted QRD) in the mid-1970s was based on the number theory that is employed by RPG Diffusor Systems, Inc.

Sound Propagation Indoors

Whenever a sound source is enclosed inside a room, the room dimensions and shape, as well as the reflective properties of the terminating surfaces (walls, floors, and ceiling) and any other obstructions to the path of sound travel, affect the sound level and quality within the space. The principal acoustic phenomena that occur within rooms are reverberation, echoes, sound concentrations, and room resonance.

FIGURE 1-10.   A two-dimensional diffusor panel to spread sound in a hemispherical pattern.   (Courtesy of RPG Diffusor Systems, Inc., 651-C Commerce Drive, Upper Marlboro, MD 20772. With permission.)

Reverberation

When the terminating surfaces (i.e., walls, floors, and ceilings) within a room are acoustically reflective, sound waves bounce off of these surfaces, causing a buildup of sound energy such that sound levels are independent of, or change very slightly with, location within the room. All sound sources within the room would blend together so that speech would be difficult if not impossible to understand. The combination of all sounds within a room having reflective terminations is known as reverberation. Reverberation can amplify sounds within a room in addition to minimizing their intelligibility. This can cause noise problems that can be alleviated by adjusting the room shape and absorption characteristics. Reverberation is not always undesirable. Some reverberation is desirable in music spaces to provide blending of the sounds to the audience. Reverberation parameters and control are discussed in Chapter 4.

Echoes

The human brain processes sound signals in such a way that, if the difference in the arrival time between two separate, similar sounds is any greater than 0.05 (1/20) s, we would hear two distinct sounds; if it is any less than 0.05 s, we would hear only one sound. At average room temperatures (70°F), during this time period a sound wave would travel around 60 ft, because the speed of sound (in units of feet per second) multiplied by time (in units of seconds) results in distance (in units of feet). If a reflective surface is anywhere within a room, there would be at least two paths for a sound wave to travel to arrive at the same listener. One would be that traveling in a line directly from the sound source to the listener. This would be the shortest distance between them. The other path would be from the source to the reflective surface, bouncing off the reflective surface, and traveling to the listener. The latter sound path distance would obviously be longer than the former. From our discussion above, if the difference between these two sound paths is greater than 60 ft, the listener would hear two distinct sounds, delayed in time, from the single source. This phenomenon is known as an echo. Echo generation is illustrated for clarification in Fig. 1-11.

A special type of echo that occurs when a sound source is between parallel reflective walls is known as flutter echo. In this case the sound bounces back and forth between the walls many times, creating a flapping noise that sounds like bats or birds are flying in the area. This usually occurs only when the parallel walls are tall and fairly close to each other.

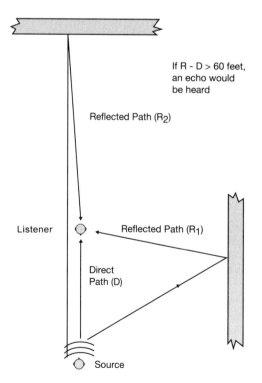

If R - D > 60 feet, an echo would be heard

Reflected Path (R$_2$)

Listener

Reflected Path (R$_1$)

Direct Path (D)

Source

FIGURE 1-11.   Echo generation.

Sound Concentrations

A concave reflective surface can have the effect of an acoustical lens to focus many sound waves at a specific location. In this case, sounds can be loud in one area and inaudible in others. The results of this focusing are known as sound concentrations. Examples of the effects of sound concentrations are evident in buildings that have domed ceilings. Inside such buildings, whispers may be heard from people on opposite sides of the room while shouts may be barely audible from other locations closer to the listener. Figure 1-12 illustrates this phenomenon.

Room Resonance

In addition to flutter echo, a sound source between two parallel reflective walls can set up a condition known as room resonance (or acoustic resonance, as opposed to the mechanical type of resonance described briefly in the discussion on infrasound) at specific frequencies. These frequencies can be estimated by solving the following equation:

$$f_n = nc/2D \qquad (1\text{-}4)$$

where $n = 1, 2, 3, \ldots$ (any positive whole number), c is the speed of sound, and D is the distance between reflective surfaces.

This is illustrated conceptually in Fig. 1-13. f_1 is known as the fundamental frequency, and f_2 and f_3, are called the second and third harmonics, respectively, of the room. Harmonics are therefore integer multiples of the fundamental frequency, where the fundamental frequency corresponds to one-half the wavelength of the sound.

Room resonances, or modes, exist in all three dimensions of a room. In a rectangular room, the simplest form of resonance would occur between two parallel walls, as discussed above. This is an example of what is known

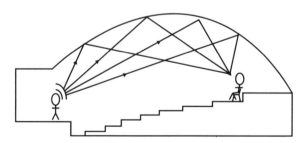

FIGURE 1-12.   Generation of sound concentrations by concave surfaces.

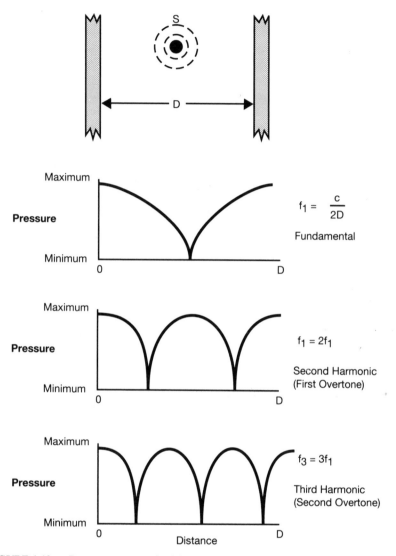

FIGURE 1-13.   Pressure patterns of axial room modes.

as an axial mode. Modes in two dimensions, parallel to a floor and ceiling and reflecting from each of the four walls of a rectangular room, are called tangential modes. Those in three-dimensional space, reflecting off the walls, floor, and ceiling, are called oblique modes. These modes cause uneven sound distribution within a room. They are of particular concern in rooms where

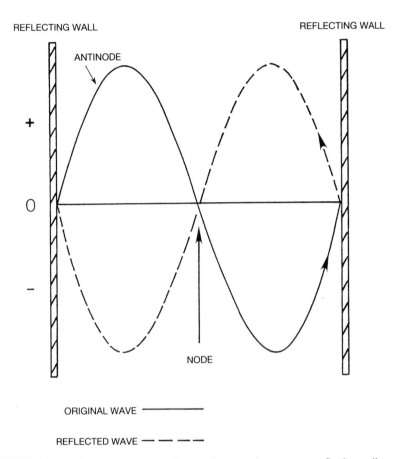

FIGURE 1-14.   The pressure pattern of a standing wave between two reflecting walls.
—, Original wave; ———, reflected wave.

music or speech are to be appreciated, because of the distortion they generate. In terms of noise, room modes can amplify sounds that have dominant frequency components matching a room resonance frequency.

The condition of room resonance is caused by the reflection of an acoustic wave, at a specific frequency, and the resultant interaction between the incident and reflected waves, as illustrated in Fig. 1-13. At integer (whole number) multiples of half-wavelengths, sound cancellations (occurring at nodes) and reinforcements (occurring at antinodes) occur because half of the sine wave is opposite in phase (pressure pattern) to the other half of the wave. The sound then reflects out of phase with the incident wave and the resulting pressure pattern is shown in Fig. 1-14. The pressure pattern resulting from this interaction stands still within the room at the room resonance frequencies, thus giving the name of standing waves to this occurrence.

The discussion above is based on the conditions of two parallel, fixed reflective surfaces. Acoustic resonance can also occur when one or both space boundaries consist of sharp discontinuities in cross-sectional areas. An example of this condition is an organ pipe, which has one end fixed and solid while the other is open to the outside air. In such a case, the acoustic resonance frequencies are different from those described above and are defined by the following:

$$f_n = (2n - 1)c/4D \qquad (1\text{-}5)$$

where $n = 1, 2, 3, \ldots$ (any positive whole number), c is the speed of sound, and D is the length of the pipe. This means that the fundamental frequency of a system open at one end and closed at the other corresponds to one-quarter of the sound wavelength, f_2 corresponds to three-quarters of the wavelength, and so forth. Figure 1-15 illustrates this form of acoustic resonance.

Sound Propagation Outdoors

As was mentioned in the section Refraction (above), sound propagation direction changes with changes in atmospheric conditions. In terms of temperature variations alone, sound waves tend to bend toward cooler temperatures. An example of wave bending caused by temperature variations would be sound travel on a typical summer afternoon. In this case, air temperatures generally decrease with altitude and sound generated at ground level would bend upward toward the cooler air. When this occurs, a source may be visible at a distance but quieter than expected, if audible at all. The sound waves are bending up and over the person listening, creating what is known as a shadow zone. The other extreme would occur when the air is cooler close to the ground than it is at higher altitudes. This happens late at night or over calm lakes or icy surfaces. In these cases, sound waves bend downward toward the ground and, if the ground is reflective, the sound bounces off the ground and hops along to propagate much farther than expected. Because still water is highly reflective, quiet conversations may be heard from opposite ends of a lake as a result of this phenomenon. These extremes of sound propagation through temperature changes are illustrated in Fig. 1-16.

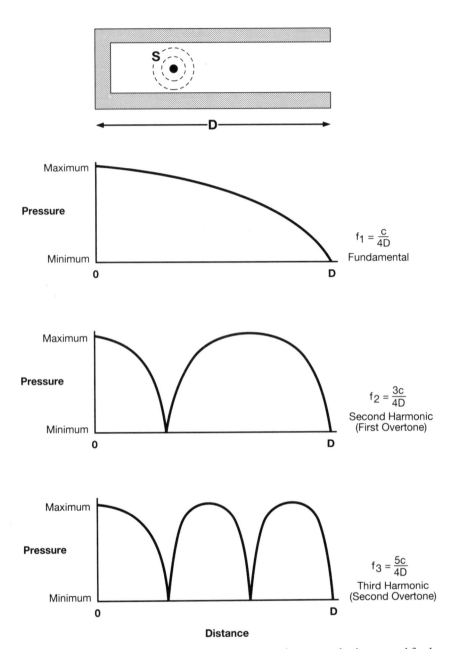

FIGURE 1-15.   Pressure patterns of acoustic resonance in a system having one end fixed and one end open.

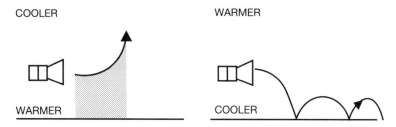

FIGURE 1-16.   The effects of temperature variations on sound propagation direction. The shaded area is the shadow zone.

In general, wind currents allow sound to be propagated farther than expected when the source is emitting sound in the direction of wind travel (downwind), and less than expected when the source is emitting sound in the direction against the wind (upwind). A combination of wind and temperature effects is responsible for the common occurrence of aircraft noise fading in and out of hearing range while the aircraft is traveling toward the listener. The effects of wind currents are illustrated in Fig. 1-17.

As sound travels away from its source, its energy is dissipated by atmospheric absorption, by obstructive surfaces (ground, vegetation, and structures), and by the spreading of the sound wave over an increasingly larger area. The specific amount of sound energy reduction with distance from the source is discussed in Chapter 3.

Echoes and reverberation can occur outdoors when reflective surfaces are close enough to the sound source that the reflected sound energy is not dissipated into the atmosphere. The urban canyon effect of tall city buildings lining streets can contribute to these effects.

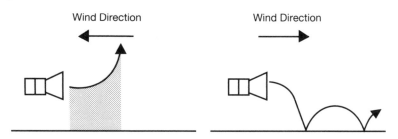

FIGURE 1-17.   The effects of wind currents on sound propagation direction. The shaded area is the shadow zone.

THE HEARING MECHANISM

A discussion of environmental noise pollution would not be complete without discussing the general operation of the human hearing mechanism. This is the ultimate receiver of sounds. Its operation is the reason for all studies and efforts in this discipline. Figure 1-18 presents the human hearing apparatus broken into the commonly designated segments of the outer, middle, and inner ears. The path of the sound wave through this mechanism can be visualized by observing Fig. 1-18 while reading the discussion below.

How We Hear

Outer Ear
Sound first impinges on the pinna, which serves as a horn to channel the sound and to smooth the transition from the outside air to the ear canal. Whenever a sharp transition in cross-sectional area occurs, as is the case

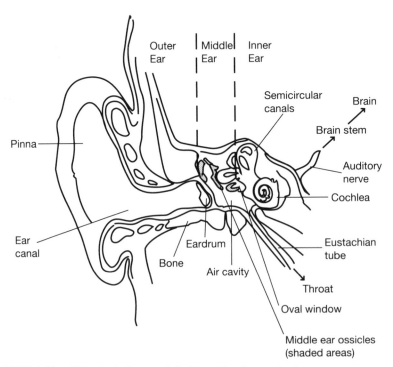

FIGURE 1-18.   The principal parts of the human hearing mechanism.

when sound travels from the outside air to a small tube, there is what is known as an impedance mismatch between the air and the tube. Acoustic impedance, or the ability of the system to impede or allow the flow of acoustic energy, is dependent on the cross-sectional area of the sound path. When there is a sharp change in cross-sectional area in the sound travel path, there is also a sharp change in acoustic impedance, which causes a restriction of sound energy flow. A gradual cross-sectional area change, afforded by a horn-shaped device, minimizes this acoustic flow restriction and allows most of the sound energy to travel from the outside air to the ear canal for our hearing reception. This is the same principle that is behind the design of horns for loudspeakers. Without the horns, the sound quality emitted by loudspeakers would be significantly degraded.

The pinna thus focuses sound waves into the ear canal. The length of the average ear canal is 3 cm or 1.2 in. The ear canal, because it is open to the air at one end and closed off by the eardrum at the other end, operates as the organ pipe system referred to earlier. This sets up a fundamental acoustic resonance between 2700 and 3500 Hz (depending on temperature and ear canal size), using Eq. (1-5). This accounts for the amplification in human hearing frequency sensitivity mentioned at the beginning of this chapter. Ear wax is secreted by the walls of the ear canal to collect dirt and impurities that would otherwise harm the eardrum. The downward tilt of the canal combined with gravity forces the wax down and out of the ear canal. Ear wax, then, serves a protective function and should be washed out only with products specifically designed for that purpose. Most people use cotton swabs for this purpose. Because most of these swabs are too big to remove the wax, they often compact the wax against the eardrum. This can cause infection and blockages that could affect hearing sensitivity. Using smaller, sharp objects can puncture the eardrum, and therefore they should also be avoided.

Middle Ear

Once the sound traveling down the ear canal reaches the eardrum, known clinically as the tympanic membrane, the eardrum vibrates in a pattern similar to that of the acoustic pressure wave. The vibrating eardrum causes the middle ear ossicles to vibrate, thus carrying the signal to the inner ear through the oval window at the entrance to the cochlea. The middle ear ossicles, known as the malleus (or hammer), incus (or anvil), and the stapes (or stirrup), are tiny bones enclosed in an air cavity sealed off by bone, the eardrum, cochlea, and eustachian tube. The ossicles are named from their resemblance to the associated common products. They not only carry the acoustic signal to the inner ear but act as a transformer to amplify the signal more than 20 times between the eardrum and the oval window. This amplification is necessary because the sound energy is significantly reduced when it passes from the air to the fluid in the inner ear. The amplification afforded by the ossicles compensates for this loss of energy in the middle ear-to-inner ear transfer.

The only connection between the outside air and the middle ear air cavity is the eustachian tube, which connects with the back of the throat. This connection is open to the outside air only when a person swallows or yawns. This explains why pressure is felt on the eardrum when we experience sudden changes in altitudes in airplanes or elevators. The atmospheric pressure outside the head changes with altitude but, because the middle ear is sealed off from the outside world, the pressure in the middle ear does not change and the pressure differential thus created causes pressure on the eardrum and thus discomfort in our ears. When one swallows or yawns, the eustachian tube is opened to the throat and the pressure is allowed to equalize with that of the outside environment, relieving the pressure buildup inside the middle ear. Babies should be fed while descending or ascending in airplanes to avoid the discomfort of this pressure buildup.

Another common middle ear-related problem with babies and young children is middle ear infection, clinically known as otitis media. This occurs more frequently in young children than adults because babies are usually born with their eustachian tubes tilted downward from their throats to their middle ears. Any throat or nose infection would then have easy access to the middle ear. As the child's head grows, the eustachian tube becomes tilted in the opposite direction, as shown in Fig. 1-18. In the adult configuration of the eustachian tube, only severe illnesses of the nose and throat can affect the middle ear cavity.

In addition to the three bones that are suspended in the air cavity by ligaments, the middle ear contains two muscles—the tensor tympani muscle and the stapedius muscle. The tensor tympani muscle connects between the malleus and the bone wall of the middle ear near the eustachian tube, and the stapedius muscle connects the stapes to the bone cavity. Both muscles connect to their respective bones through tendons. These muscles contract when the ear is exposed to high sound levels. This muscle contraction, clinically known as acoustic reflex, moves the bones with respect to one another to reduce sound level transmission to the inner ear, protecting it from potential damage. These muscles take some time to react and lose their contraction ability with time. Therefore, this protection is not effective on sudden impulsive noise exposures or long-term exposures. As the sound reduction ability of acoustic reflex is also limited, high-level noises can still reach the inner ear to cause significant damage.

Inner Ear

Once the sound wave reaches the oval window, it causes the fluid inside the spiral-shaped cochlea to vibrate. The cochlea is encased in bone and lined with millions of tiny, flexible hair cells that wave with the vibrations set up in the fluid at the oval window. The hair cells are actually nerve cells connected to the auditory nerve. The location of the hair cells along the cochlear spiral dictates the frequency sensitivity of the hair cells. The higher frequency hair cells are closest to the oval window, as would follow from their shortest wavelength. The fluid vibrations in the cochlea are similar to ocean waves rolling onto a coastline. The farther up the cochlea these waves land, the lower the frequency that is interpreted, because the longer wavelengths represent the lower frequencies. Each stimulated hair cell sends an electrical signal to the auditory nerve. The actual frequency interpretation takes place after the stimulated hair cells send their respective electrical signals through the auditory nerve to the brain. The exact process of this interpretation is still being studied by many research institutions and audiologists.

Bone Conduction

Sounds can also be sensed by bone conduction, although to a much lesser extent than by conduction through the pinna. This occurs, principally for frequencies below 1000 Hz, when the vibrations of the bones in the skull are transmitted to the ear canal and middle ear ossicles. Bone conduction makes your voice sound different to you than to another person, explaining why most people say that their voice sounds different to them when it is played back from a recording. Bone conduction becomes significant to hearing when the pinna is covered or when the ear canal is sealed off by a hearing protective device.

How Hearing Loss Occurs

When discussing hearing loss in the context of environmental noise pollution, it is often assumed that most hearing loss results from exposures to high levels of noise. Although many hearing loss cases are caused by exposures to high noise levels, there are also many other possible causes for hearing loss. Among these, the most common are heredity, disease, certain medications, aging, and physical damage to components of the hearing mechanism. A diagnosis of noise-induced hearing loss should not be established until these other possible causes are ruled out.

There are generally two classes of hearing loss: conductive and sensori-neural. Conductive hearing loss results from physical abnormalities or blockages in the outer and middle ears and is reparable in most nonsevere cases. Sensorineural hearing loss results from nerve damage in the inner ear and auditory nerve and is usually irreparable. Noise-induced hearing loss is sensorineural.

Noise-Induced Hearing Loss

There are generally two types of noise-induced hearing loss, one that can cause permanent immediate damage (known as acoustic trauma) and one that occurs gradually over the course of years. Acoustic trauma can result from exposures to extremely high levels of sound. In this case the inner ear tissues that support the cochlear hair cells can literally be torn apart from being stretched beyond their limits. The most common sound sources that produce levels high enough to cause acoustic trauma are guns or other mechanisms that cause explosive events.

When high-level noise exposures are continuous and below the limits of acoustic trauma, cochlear hair cells are continually overstimulated to the point at which they lose their resiliency and die off. This occurrence would result in permanent hearing loss in the frequency range to which the dead hair cells had originally responded. This typically first occurs in the 2000- to 4000-Hz range, where our hearing mechanisms have the greatest sensitivity. A warning signal that these hair cells have been overstimulated is a ringing or buzzing noise, clinically known as tinnitus, heard after the high-level noise exposure has stopped. It must be noted, however, that tinnitus has many potential causes beside high-level noise exposure.

Tinnitus is commonly caused by overstimulation of the auditory nerve. Just as pinching the skin registers a sensation of pain in the brain by stimulating nerves near the skin surface, stimulating the auditory nerve by pressure, swelling, or noise exposure registers as the sensation of extra sound in the brain. Tinnitus, when caused by high-level noise exposure, can take anywhere from minutes to days to subside, but normally subsides within a few hours of the high-level exposure. If the tinnitus has not subsided after a few days, medical consultation is suggested.

Normal hearing is usually restored after the hair cells stabilize, but permanent hearing loss can result from repeated exposure. Because noise-induced hearing loss usually occurs gradually, people normally do not notice it until a significant loss has already occurred. It is common for people exposed to high noise levels to say that they have become accustomed to the noise that once bothered them. What has usually occurred, however, is that they have lost enough of their hearing sensitivity that the noise has stopped bothering them. Another common misconception about noise-induced hearing loss is that it is somehow related to physical strength or endurance. Noise-induced hearing loss is caused by nerve damage and has nothing to do with physical strength. At this juncture in medical history, this nerve damage is irreparable.

Noise-induced hearing loss usually begins with a drop in sensitivity to sounds in the 2000- to 4000-Hz range, where many consonant sounds lie. The person with this type of hearing loss, then, would not hear consonants clearly and has difficulty understanding speech. The vowel sounds would come through clearly, so raising the voice level of the speaker would make the voice sound louder to this person without improving intelligibility significantly.

Noise-induced hearing loss is usually divided into two categories: environmental (known as sociocusis) and occupational. Occupational noise exposure is discussed briefly in Chapter 5 but environmental sources and noise effects are the principal topics covered in this book. Specific noise-induced hearing loss risk criteria and noise effects are discussed in Chapters 2, 5, and 6.

Heredity

Under the category of heredity would be otosclerosis, predisposition to early degeneration of the auditory nerve, and anatomical malformations. Otosclerosis is a disorder that causes a hardening (or sclerosis) of the connection between the stapes and the oval window. This hardening results in hearing loss because the stapes cannot effectively transmit the sound signal to the inner ear with this condition. Otosclerosis is normally treated by surgically removing the stapes (known as stapedectomy) and replacing it with a wire connecting the incus to the oval window. This wire transmits the acoustic signal from the incus to the oval window, a task the stapes could not accomplish because of its rigid connection to the oval window.

Predisposition to early degeneration of the auditory nerve usually becomes most obvious in children from birth to 6 years of age. At this time, there is no known cause for this condition. Anatomical malformations can vary from slight defects to the complete absence of the entire hearing mechanism.

Disease

There are many diseases that have the potential to cause temporary and permanent hearing loss. Any acute illness, particularly if accompanied by a high fever, can damage cochlear nerve endings. Smoking, drinking alcoholic beverages, or drug taking during pregnancy may also produce hearing impairment in the unborn child (Glorig, 1974). Common diseases that have been known to cause hearing loss include the common cold, allergies, otitis media, syphilis, diphtheria, typhoid fever, pneumonia, influenza, mumps, measles, scarlet fever, meningitis, and encephalitis (EPA, 1981). Tumors can cause hearing loss by applying pressure to the auditory nerve and thus choking its ability to carry sound signal information to the brain.

Medications

Several antibiotics, aspirin, quinine, and nicotine have been linked to causing hearing loss (Glorig, 1974).

Aging

Presbycusis is the clinical term for the loss of hearing sensitivity with age. There is some debate that sociocusis causes presbycusis, stemming from studies such as the one performed by Rosen (1967) on a primitive African tribe living in a noise-free environment. According to this study, no presbycusis was observed in elderly men; however, other factors, such as diet and difficulty in age determination, may have influenced these findings (EPA, 1981).

It is generally agreed that presbycusis is caused by degeneration of the nerves and circulatory system involved in the hearing mechanism. The effects of presbycusis may appear in people as young as 20 years of age, usually beginning in the 2000- to 4000-Hz range (the range of our greatest sensitivity to sound). Specific values of hearing loss associated with presbycusis are given in Chapter 2.

Physical Damage

Included in the category of physical damage would be impacted ear wax clogging the ear canal, eardrum rupture, injury caused by insertion of foreign objects into the ear canal, concussion, skull fracture, any accidental occurrence, and surgical interference.

REVIEW QUESTIONS

For all calculations, assume a temperature of 70°F unless mentioned otherwise.

1. Come up with a common illustration for frequency and wavelength other than the two mentioned in text (water and spring).
2. To what frequency range are our hearing mechanisms most sensitive, and why? What is the dominant range of human speech frequencies?
3. What is a warning signal that tells you that you have been exposed to noise levels that are potentially hazardous to your hearing? How can you tell if you have the first signs of noise-induced hearing loss?

4. Rewrite Table 1-1, replacing the speed of sound column with wavelength for a 1000-Hz signal.
5. Calculate the percent change in speed of sound in air with temperature variations (i.e., percent change per 10°F).
6. What are the four principal ways that sound changes direction of propagation when interacting with a change in medium or medium condition?
7. a. If air temperature decreases with altitude and a sound wave were directed upward from the ground at an angle of 45°, in what direction would the wave bend? Draw a trace of the acoustic wave travel direction in the atmosphere.
   b. How would the same wave travel if temperature increased with altitude?
8. a. A rectangular room bordered by four smooth, flat walls has floor dimensions of 15 by 25 ft. Calculate the fundamental and second and third harmonic frequencies for axial standing waves in the room.
   b. Calculate the fundamental resonance frequency of the human ear canal, assuming its length is 3 cm (2.54 cm = 1 in.) and the temperature is 65°F.
9. Why does your voice sound different to you than it does to others?
10. Name at least 10 possible causes for hearing loss and the class of loss (conductive or sensorineural) involved in each case.

References

EPA. 1981. *Noise Effects Handbook*. EPA 550/9-82-106. Washington, D.C.: U.S. Environmental Protection Agency.

Glorig, A. 1974. *A Doctor Talks about Hearing Loss*. Minneapolis, MN: Qualitone, Inc.

Rosen, S. 1967. Presbycusis study of a relatively noise-free population in the Sudan. Ann. Otol. Rhinol. Laryngol., 727–743.

2

Noise Descriptors

SOUND LEVELS

The range of acoustic pressures to which our hearing mechanisms respond is from 2×10^{-5} newtons per square meter (N/m^2) at the threshold of hearing to $20\ N/m^2$ at the threshold of pain, a ratio of 1 million to 1 from one threshold to the other. Combining this vast range of pressure sensitivities with the fact that we perceive a doubling of sound power as a just noticeable change, the logarithmic scale was adopted as being the most appropriate for describing sound levels. A logarithm is an exponent of a specified base, base 10 in this case. Therefore, the logarithm to the base 10 of 100, designated as $\log_{10} 100$, is 2 ($100 = 10^2$), $\log_{10} 1000$ is 3 ($1000 = 10^3$), and so on. In typical notation, the subscript 10 is dropped and it is understood that "log" implies 10 as a base.

The other common base used for logarithms is the natural exponential function, denoted by the letter e. e is 2.7183 and the logarithm to the base e is known as the natural logarithm, denoted by ln. To convert ln to log,

$$\log(a) = 0.4343 \times \ln(a) \tag{2-1}$$

where a is any number.

For reference, the mathematical properties of all logarithms are as follows:

a. $\log(a \times b) = \log(a) + \log(b)$

b. $\quad \log(a/b) = \log(a) - \log(b) \tag{2-2}$

c. $\quad \log(a^b) = b \times \log(a)$

The term *level* implies a logarithmic ratio of some parameters. The term *sound level* implies a logarithmic ratio of sound power or parameters proportional to sound power. The basic unit of sound level is the decibel (denoted dB). It stems from the use of the unit called a bel (named after Alexander Graham Bell [1847–1922], the American inventor of the telephone), corresponding to a multiple of 10 of the threshold of hearing. Because the bel (denoted B) is too large to describe the acoustic environment adequately, the unit was broken into tenths, or decibels. Therefore, 10 dB corresponds to 1 B, and the decibel is used as the unit of choice.

Decibels

Definitions
One decibel of sound level is defined from:

$$L_W = 10 \times \log(W/W_{ref}) \tag{2-3}$$

where W (standing for watts, a unit of power) is the measured power, W_{ref} is a reference power, usually 1×10^{-12} W, and L_W is the sound power level. Sound pressure is the parameter that is normally measured in noise assessments. As sound pressure squared is proportional to sound power, it can be denoted in terms of decibels. The resulting quantity is known as the sound pressure level (SPL), and is defined as

$$SPL = 10 \times \log(p^2/p_{ref}^2) \tag{2-4}$$

where p is the measured acoustic pressure and p_{ref} is the reference pressure of 2×10^{-5} N/m^2, 20 μPa (micropascals), or 2×10^{-4} μbars (depending on the system of units employed), corresponding to the threshold of hearing. Using Eq. (2-2c), the SPL definition above can be rewritten as

$$SPL = 20 \times \log(p/p_{ref}) \tag{2-5}$$

Because the log of 1 is 0 ($10^0 = 1$), the SPL = 0 dB when the acoustic pressure is the same as the threshold of hearing. Therefore, 0 dB corresponds to the threshold of hearing, or the SPL at which people with healthy hearing mechanisms can just begin to hear a sound.

Decibel Mathematics
When adding sound levels from two sources, logarithmic mathematics must be used. Table 2-1 helps provide accurate results without having to deal with the mathematics. With Table 2-1, it can be seen that 60 dB + 60 dB would

TABLE 2-1   Adding Decibels

Difference between the Two Levels Being Added (dB)	Decibels to Add to the Higher Level
0 or 1	3
2 to 4	2
5 to 9	1
10 or more	0

not equal 120 dB, as arithmetic mathematics dictates, but 63 dB, as logarithmic mathematics dictates.

Note from Table 2-1, that if two sound sources differ in level (at the same location) by more than 9 dB, the addition of the two source levels will result in the same level as that with the louder source alone, and it can be said that the louder source dominates the acoustic signature of the noise environment. It also shows that if two sources emit the same sound levels separately and are both operated simultaneously (or if the power of a single sound source were doubled), the sound level would increase by 3 dB. This rule would hold true for any decibel values, even 0 dB, because 0 dB corresponds to a definite amount of pressure, power, or whatever unit is used in the decibel. Therefore 0 dB + 0 dB = 3 dB.

If many identical sound sources were operated simultaneously at the same location, the total sound level would be the level of sound from one of the sources plus $10 \times \log(n)$, where n is the number of identical sources. There-

TABLE 2-2   Multiple Source Corrections

Number of Identical Sources	Decibels Added to Single Source Value
2	3
3	5
4	6
5	7
6	8
7	8
8	9
10	10
15	12
20	13
50	17
100	20

fore, if 1 source was measured at 50 ft to generate a 60-dB SPL, 10 identical sources operating at the same location would generate a 70-dB SPL and it would take 100 of these sources operating at the same location to generate an 80-dB SPL. For reference, Table 2-2 is included with factors corresponding to $10 \times \log(n)$.

Note from Table 2-2 that each time the number of sources is doubled, 3 dB is added to the sound level, and each time the number of sources is changed by a factor of 10, 10 dB is added to the sound level.

When adding more than two different sound levels, add two levels at a time and add the result to the third, that result to the fourth, and so on. Refer to Table 2-1 to verify the following decibel additions. In example (e) below, using either Table 2-1 or 2-2 should provide the same answer.

a. 70 dB + 70 dB = 73 dB

b. 70 dB + 80 dB = 80 dB

c. 70 dB + 76 dB = 77 dB

d. 70 dB + 71 dB + 72 dB + 73 dB = 78 dB

e. 70 dB + 70 dB + 70 dB + 70 dB = 76 dB

When one needs to know what the sound level of a contributing source would be by itself, without other sources present, decibel subtraction can be performed. This would be necessary if several sources are operating simultaneously and one needs to be isolated, or if the source level does not dominate (more than 9 dB over) the background noise level. Table 2-3 can be used

TABLE 2-3   Subtracting Decibels

Difference (in dB) between SPL with All Sources Operating and SPL with Source(s) of Interest Not Operating	Decibels to Be Subtracted from SPL with All Sources Operating to Derive SPL from Source of Interest Alone
0	At least 10
1	7
2	4
3	3
4 or 5	2
6 to 9	1
10 or more	0

for this purpose on a general level. More specific correction factors, as recommended by published standards, are listed in Chapter 3.

As an example of decibel subtraction, let us consider a situation in which the SPL with a machine operating is 70 dB at 50 ft and the background noise level (without the machine operating) is 65 dB at the same location. The difference between the two levels is therefore 5 dB. With Table 2-3, this would mean that the SPL generated by the machine alone would be 70 − 2 or 68 dB. Going backward in this calculation, 65 dB + 68 dB = 70 dB, Table 2-1 values may be used to verify the results.

Note the distinction between ambient and background levels. These two are not the same although many people use these terms interchangeably. The ambient level includes all sounds in the area, including the noise source of interest, whereas the background level includes all sounds except the noise source of interest.

Common Misconceptions

It should be noted that sound levels must be qualified in terms of their ratioed parameters for the level to have meaning. As is stated above, a decibel-level quotation tells the reader that a logarithmic ratio is being described; however, that could be a ratio of powers, pressures, intensities, voltages, accelerations, or any quantity that the author chooses. Two common ways of qualifying decibel levels are by specifically naming the level designation (e.g., sound pressure level implies a pressure ratio, sound power level implies a power ratio) or by stating the ratio reference after the decibel notation (e.g., dB re 2×10^{-5} N/m^2 implies sound pressure level because the reference units are those of pressure; dB re 1×10^{-12} W implies sound power level because the reference units are those of power). Without these qualifications, a decibel quotation is meaningless.

Another point to bear in mind when dealing specifically with SPL quotations is that, because acoustic pressure varies with distance from the source, so does the SPL. Therefore, an SPL quotation must also include the location with respect to the source to have meaning. Because sound power is independent of location, this specification is not needed for sound power levels; but it must be given for SPL quotations.

In general, human sound perception is such that a change in SPL of 3 dB is just noticeable, a change of 5 dB is clearly noticeable, and a change of 10 dB is perceived as a doubling or halving of sound level.

If two sound signals are combined that are identical in frequency and phase (their pressure patterns are identical in space and time), adding the two signals would double the pressure and, according to Eq. (2-5), would cause a 6 dB increase in the signal monitored from only one of the sources. In this case, the two sources would be considered to be coherent. Under typical circumstances, however, coherent sources are rare because, even if two sources were nearly identical, it would be highly unlikely that their phase characteristics would be perfectly aligned. Therefore most noise sources involved in environmental assessments are considered to be incoherent. A more realistic estimate of combining incoherent sources having identical SPL values is to add 3 dB to the SPL monitored from one source alone, as would be the case when sound power is doubled (from Eq. (2-3)). This makes sense when one thinks in terms of doubling the number of sources, implying doubling the power of the original source.

In a large open area with no obstructive or reflective surfaces, beginning at least 15 ft away from a sound source that is small in size compared to measurement distances, it is a general rule that the SPL drops off at a rate of 6 dB with each doubling of distance away from the source. This rule is known as the inverse square law, named for its mathematical basis that sound intensity decreases at a rate inversely proportional to the square of the distance from the source. For example, if an outdoor source in an open area is known to produce an SPL of 80 dB 50 ft away, it would be expected to measure at 74 dB 100 ft away and 68 dB 200 ft away. Beginning 300 to 500 ft from the source, atmospheric attenuation of the sound must also be taken into account. Specific atmospheric attenuation factors are listed in Chapter 3. Also, the drop-off rate will vary with terrain conditions and where obstructions exist in the sound propagation path. Therefore, in the urban canyon-type environment existing in many large cities, the inverse square law cannot be used. Because of sound reflections off buildings, drop-off rates along streets in downtown areas of cities generally range from 2 to 4 dB per doubling of distance from the source. Whenever ideal open situations do not exist and a drop-off rate is required in the analysis, the rate should be verified by measurements.

If the sound source is composed of many individual sources spread out along a line, it is called a line source. The inverse square law is based on the spherical spreading of sound energy from a point in space, known as a point source, having dimensions that are small compared to monitored distances. Sound radiating from a line source would spread in a cylindrical pattern. In this case the SPL would drop off at a rate of 3 dB with each doubling of distance from the source. Typical line sources include traffic on highways, and long trains.

Measurement Parameters

There are many descriptors commonly used in environmental noise assessment. The choice of specific descriptors is related to the nature of the noise signature of the source and the potential effect it may have on the surrounding environment. Not only would different descriptors be most appropriate for different types of sources, but different measurement techniques may also be appropriate. The most common types of measurement used in the field of environmental noise assessment are overall and octave band measurements. Most noise measurements are in terms of SPL. All discussions about sound levels in this document will refer to the SPL unless specified otherwise.

Weighting Networks

An overall measurement results in a single decibel value that describes the sound environment, taking all frequencies into account. Most basic sound level meters measure sound in terms of overall sound pressure levels, using different so-called weighting networks. The most common weighting networks used are the A- and C-weighting networks. Weighting networks are filters in sound level meters that vary frequency sensitivities according to set standards. Because the human ear does not sense all frequencies in the same manner, we do not hear the sounds the same way a typical microphone would. Over the normal hearing range of 20 Hz to 20 kHz, we are most sensitive to sounds with frequencies between 200 Hz and 10 kHz. Most microphones do not attenuate or amplify sounds in this frequency range the way our hearing mechanisms do. The A-weighted scale was developed as a set of filters in sound level meters that simulate the frequency sensitivity of the human hearing mechanism. Because human reaction is normally the reason for an environmental noise assessment, A-weighted decibels (symbolized in units of dBA or dB(A)) are usually the units of choice.

Common noise sources with their associated typical dBA values are shown in Table 2-4. Note that 0 dBA corresponds to the threshold of hearing and 120 dBA corresponds to the threshold of pain, the level at which physical pain can be felt in the ears as a result of exposure to the sound.

The C-weighted network provides the unweighted microphone sensitivity over the frequency range of maximum human sensitivity. C-weighted decibels (denoted in units of dBC or dB(C)) are used in some ordinances and standards, usually when dealing with stationary mechanical noise sources; however, dBA units are normally used for environmental assessments. Because C-weighting does not attenuate frequency levels below 1000 Hz the way A-weighting does, inspection of dBA vs. dBC readings can give the inspector a quick estimate of the low-frequency contribution of the sound source in question. For example, if the dBC reading is much higher than the dBA reading, this would be an indication of a strong low-frequency dominance in the acoustic signature. On the other hand, if the dBA and dBC levels are comparable (to within 3 dB), the source dominance is in the higher frequency range. A- and C-weighted standard frequency sensitivities are described numerically in Chapter 3.

TABLE 2-4 **Noise Levels of Common Sources**[a]

Sound Source	SPL (dBA)
Air raid siren at 50 ft (threshold of pain)	120
Maximum levels in audience at rock concerts	110
On platform by passing subway train	100
On sidewalk by passing heavy truck or bus	90
On sidewalk by typical highway	80
On sidewalk by passing automobiles with mufflers	70
Typical urban area background/busy office	60
Typical suburban area background	50
Quiet suburban area at night	40
Typical rural area at night	30
Isolated broadcast studio	20
Audiometric (hearing testing) booth	10
Threshold of hearing without hearing damage	0

[a]A change in 3 dBA is a just noticeable change in SPL; a change in 10 dBa is perceived as a doubling or halving in SPL.

It is possible for A-weighted readings to be 1 or 2 dBA higher than C-weighted readings in the case of dominance in the 2000- to 4000-Hz range (accounting for ear canal amplification in this frequency range by the A-weighting network); however, a clear error in data can be spotted whenever A-weighted levels are reported to exceed C-weighted levels by more than 3 dBA. No weighting network applies signal adjustment at 1000 Hz, and therefore readings at 1000 Hz should be identical (whether weighted or not).

It should be noted that the A-weighting network is based on human hearing sensitivity to SPLs below 70 dB. Especially for SPLs greater than 90 dB, our frequency sensitivities level off at lower frequencies such that C-weighting may be more appropriate to quantify human sound perception. Most agencies and rating methods, however, use A-weighting exclusively for noise assessment, independent of SPL, for ease of data taking and consistency of results.

Frequency Bands

When dealing with mechanical equipment noise assessment, and especially noise control, it may become necessary to view the frequency components of the noise signature. In these cases, overall readings do not provide enough information and frequency band analysis, known as spectrum analysis, must be used. The term *spectrum* refers to a graph of sound level plotted against frequency, the characteristic plot of the frequency content of a sound source.

The most common type of spectrum analysis is known as octave band analysis.

In octave band analysis, the frequency spectrum is divided into sections (called bands) on a logarithmic basis. The characteristics of octave bands are that each band is identified by a so-called center frequency that is not in the arithmetic, but the logarithmic, center of the band; that the frequency range of each band is such that the upper limit is twice the lower limit; and that each successive center frequency is twice its preceding center frequency. The commonly used center frequencies in octave band analysis are 63, 125, 250, 500, 1000, 2000, 4000, and 8000 Hz. These frequencies are standardized internationally through the International Electrotechnical Commission (IEC; Geneva, Switzerland), and are established on a national level by the American National Standards Institute (ANSI) under ANSI Standard S1.6-1984, *American National Standard Preferred Frequencies, Frequency Levels, and Band Numbers for Acoustical Measurements.*

When more detail of the frequency spectrum is required, 1/3-octave analysis is available, which divides each octave band into three logarithmically centered bands. Specific center frequencies for 1/3-octave bands are defined by $10^{0.1N}$, where N is a whole number from 10 to 50. Depending on the required accuracy, spectrum analyzers are available that provide frequency bands as narrow as fractions of 1 Hz.

ENVIRONMENTAL NOISE DESCRIPTORS

The most common descriptors used in environmental noise assessments are L_{eq}, L_{dn}, L_n, SEL, and maximum instantaneous SPLs. Each is measured in dBA units and is described briefly below.

Definitions and Uses

The L_{eq} is the continuous equivalent sound level, defined as the single SPL that, if constant over the stated measurement period, would contain the same sound energy as the actual monitored sound that is fluctuating in level over the measurement period. The L_{eq} is an energy-average quantity that must be contrasted with an average or median sound level. The L_{eq} must be qualified in terms of a time period to have meaning. Representation of the time period (in hours) is normally accomplished by placing it in parentheses, for example, $L_{eq(1)}$ refers to a 1-h measurement and $L_{eq(24)}$ refers to a 24-h measurement. The L_{eq} is recognized as the descriptor of choice by the Federal Highway Administration for traffic sources in environmental noise assessments. In addition to its simplicity of use, it is easy to combine with other readings or predictions to derive a total noise level.

The L_{dn} (also denoted DNL) is the A-weighted day–night equivalent sound level, defined as a 24-h continuous L_{eq} with 10 dBA added to all signals recorded between the hours of 10:00 P.M. and 7:00 A.M. This 10-dBA addition is a penalty that accounts for the extra sensitivity people have to noise during typical sleeping hours. Aircraft noise around airports is usually mapped out in terms of L_{dn} contours, which are constant lines of L_{dn} mapped similarly to isobars on weather maps or elevations on topographical maps. With the exception of the Federal Highway Administration, all federal agencies having nonoccupational noise regulations use the L_{dn} as their principal noise descriptor for community assessments. These agencies include the Federal Aviation Administration, the Federal Transit Administration, the Environmental Protection Agency, the Department of Housing and Urban Development, the Department of Veterans Affairs, and the Department of Defense. In addition, ANSI Standards S12.9-1988, *American National Standard Quantities and Procedures for Description and Measurement of Environmental Sound*, Part 1, and S12.40-1990, *American National Standard Sound Level Descriptors for Determination of Compatible Land Use*, both identify the L_{dn} as the descriptor of choice for long-term environmental noise assessment measurements. Figure 2-1 provides a general indication of average outdoor L_{dn} values for different environments.

Debates have been ongoing for years between the Federal Aviation Administration (FAA) and others about the appropriateness of the L_{dn} descriptor to compensate for discrete high-level events such as aircraft flyovers. Critics say that, because the L_{dn} is a long-term descriptor and aircraft flyovers have short durations, the high noise levels associated with the short-term events would be averaged out to make the noise levels seem lower, and therefore less annoying, than they actually are. With the support of most other federal agencies, the FAA recently reaffirmed its position on this matter, that is, that the logarithmic nature of the L_{dn} causes it to stress the loudest events of its 24-h period of rating and that any arguments with this stem from a misunderstanding of the L_{dn} rating method (FICON, 1992).

The L_{Cdn} is the C-weighted day–night equivalent sound level, used when sound levels in communities have dominant high-level frequency components below 500 Hz. These low-frequency components are significantly reduced (see Chapter 3 for specific amounts) by the A-weighting network and, although the hearing mechanism reduces levels at these frequencies accordingly, sounds in this frequency range can vibrate household objects and structures, causing annoyance from the resultant rattling or buzzing. Aircraft flyovers and construction sources can generate sounds that would be appropriately analyzed by using the L_{Cdn} instead of the L_{dn}.

A variation of the L_{dn} developed in California for environmental noise assessments, is called the community noise equivalent level (denoted CNEL)

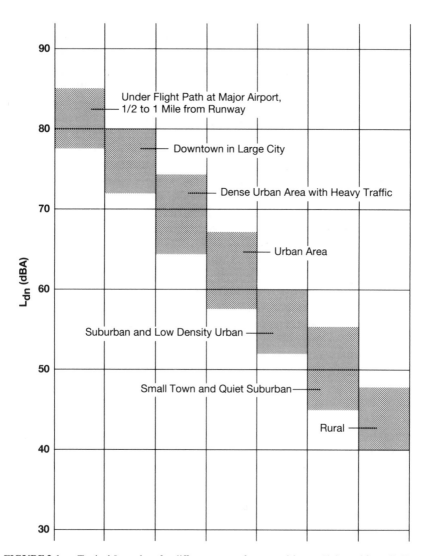

FIGURE 2-1.   Typical L_{dn} values for different types of communities.   (Adapted from DoD, 1978.)

or day–evening–night level (denoted L_{den}). For this descriptor, a 5-dBA penalty is added to measurements between 7:00 and 10:00 P.M. in addition to the 10-dBA nighttime penalty added between 10:00 P.M. and 7:00 A.M.

The L_n is the percentile level, where n is any number between 0 and 100 (noninclusive). The number designated by n corresponds to the percentage of the measurement time period by which the stated sound level has been exceeded. For example, $L_{10} = 80$ dBA means that SPL measurements exceeded 80 dBA 10% of the measurement period. As with the L_{eq}, the measurement time period must be specified and is denoted in parentheses. The most commonly quoted L_n values are L_1, L_{10}, L_{50}, and L_{90}. L_1, the SPL exceeded 1% of the time, is usually regarded as the average maximum noise level when readings are 1 h or less in duration. The L_{10} is usually regarded as an indication of truck traffic noise exposure with a steady flow of evenly spaced vehicles. The L_{50} provides an indication of the median sound level. The L_{90} is usually regarded as the residual level, or the background noise level without the source in question or discrete events.

Viewing the percentile levels provides an indication of the degree of fluctuation in noise readings. For example, a difference of more than 15 dBA between L_{10} and L_{90} values for readings of 1 h or less would indicate large fluctuations in the readings that should be verified by a high L_1 value. In these types of environments, the L_{eq} value should be close to if not exceeding the L_{10} value and the L_{90} does not provide an accurate estimate of the background noise level. The minimum level or L_{99} would provide a more realistic estimate of the background noise in this situation. Under circumstances in which noise level fluctuations are moderate (the difference between L_{10} and L_{90} values is between 5 and 15 dBA), the L_{eq} normally falls between the L_{10} and L_{50}. When the noise environment has no significant fluctuations (the difference between L_{10} and L_{90} values is less than 5), the L_{eq} should approximate the L_{50} value. These are handy checks to verify the validity of data and to describe the noise environment.

The SEL is the sound exposure level, defined as a single number rating indicating the total energy of a discrete noise-generating event, such as an aircraft flyover or train or truck passby, compressed into a 1-s time duration. Because all discrete events occur in different time durations, this level is handy as a consistent rating method that can be straightforwardly combined with other SEL and L_{eq} readings to provide a complete noise scenario for measurements and predictions. It should be noted that, because most discrete events described by the SEL last longer than 1 s and the SEL compresses all event energy into 1 s, the SEL (unless the event lasts less than 1 s) will be higher than values associated with any other rating method (including maximum instantaneous level) for a specific source.

The EPNL is the effective perceived noise level, in units of EPNdB, defined as a rating of the annoyance of a single event. The event is usually an aircraft flyover, but the EPNL can be used for any high-level noise source (e.g., trains, cars, trucks, buses, or mobile machinery) passing by an area frequented by people. Whereas all of the other descriptors referenced above can be measured directly on typical sound level meters or straightforwardly derived from such readings, the EPNL is much more cumbersome to derive. The 1/3-octave spectrum of the noise must be matched to perceived loudness curves, resulting in a modified spectrum that is further analyzed through extensive calculations that would not be appropriate to describe here. Although the EPNL is still used by the aircraft industry and by the FAA to certify their "stages" of aircraft, it is usually reserved for aircraft industry use and other descriptors are used for environmental noise assessments.

NOISE EFFECTS

The effects of noise on people have been documented for more than 30 years; however, the only effect that has an undisputed association with noise exposure is hearing loss. Other effects that have been linked to noise include stress-related illness; interference with communication, concentration, relaxation, and sleep; and annoyance. Table 2-5 lists the effects of noise exposure that have been discussed in recent literature.

Most physiological effects of noise, with the exception of hearing loss, are hypothesized to be caused by the stress associated with noise exposure. Fear reactions usually result from either not knowing the cause of the noise or feeling as if the noise source has the potential to invade or damage the structure of a shelter. Safety can be compromised by noise if levels are too high for required communication to take place. The other possible effects should be self-explanatory.

There have been several studies conducted on the effects of noise on animals (Newman and Beattie, 1985) and marine mammals (LGL, 1991), but none have been conclusive.

The principal areas of noise effects on people for which criteria have been developed are noise-induced hearing loss, speech interference, sleep interference, and annoyance.

Noise-Induced Hearing Loss

Noise-induced hearing loss criteria have been developed on a national level by the Occupational Safety and Health Administration (OSHA) for occupational risks and by the Environmental Protection Agency (EPA) for environmental risks. The OSHA prescribes a lower limit of 85 dBA $L_{eq(8)}$ and the EPA prescribes a lower limit of 70 dBA $L_{eq(24)}$ for their hearing

TABLE 2-5   Effects Linked to Noise

Effect Category	Effect
Physiological	Hearing loss
	Hypertension
	Cardiac disease
	Ulcers
	Colitis
	Endocrine and biochemical disorders
	Nausea
	Headache
	Dizziness
Psychological	Insomnia
	Annoyance
	Fear
	Stress
	Learning disability
Other	Compromising safety
	Speech interference
	Sleep interference
	Compromising privacy
	Lack of concentration
	Compromising enjoyment of leisure activities

Sources: Cherow (1991); EPA (1974, 1981); Newman and Beattie (1985); Suter (1991).

TABLE 2-6   Effects of Aging on Hearing Sensitivity to Human Speech Frequencies

	Hearing Threshold Loss (dB average of values at 500, 1000, and 2000 Hz)		
	Percent of Population		
Age (Years)	90	50	10
28	8	15	21
38	9	16	23.5
48	9	17.5	26
58	10	22	32

Source: ISO (1990). (Reprinted with permission from the American National Standards Institute.)

damage risk criteria. The specifics of these criteria are described further in Chapter 5.

On an international level the International Organization for Standardization (ISO), in ISO Standard 1999 (ISO, 1990), *Acoustics—Determination of Occupational Noise Exposure and Estimation of Noise-Induced Hearing Impairment*, established the effects of high levels of noise exposure on hearing performance. This standard also provides hearing loss values attributable to presbycusis alone, as are shown in Table 2-6. These values must be accounted for when any hearing loss evaluation is performed.

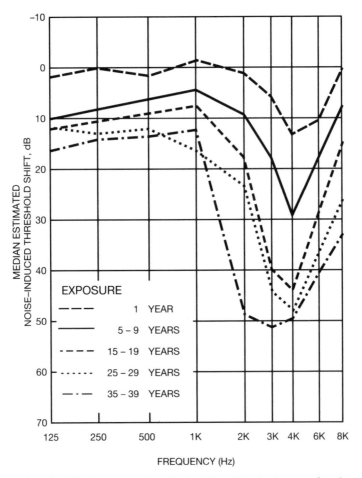

FIGURE 2-2.   Graphical representation of noise-induced hearing loss as a function of frequency.   (Data from Taylor et al., 1965; figure adapted from EPA, 1981.)

As was mentioned in Chapter 1, noise-induced hearing loss usually becomes most profound in the 2- to 4-kHz frequency range. Figure 2-2 shows the result of a sample study that illustrates this theory for industrial workers.

Speech Interference

According to the EPA, an L_{dn} of 50 dBA is the upper limit of 100% speech intelligibility both indoors and outdoors (EPA, 1974). Figure 2-3 shows the EPA-established relationship between percent intelligibility and continuous (as opposed to intermittent or impulsive) noise level. In general, ambient noise levels above 80 dBA will make speech communication impractical.

Although there are several rating methods for speech interference, the

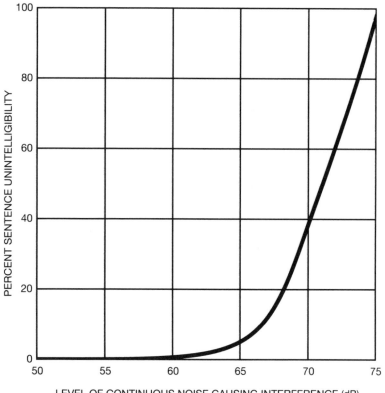

FIGURE 2-3.   Relationship between speech intelligibility and background noise for normal conversations.   (Adapted from EPA, 1974.)

Speech Interference Level (SIL) has been the one used most often for environmental assessments and is the simplest to derive. The SIL is established in ANSI Standard S3.14-1977(R1986), *American National Standard for Rating Noise with Respect to Speech Interference*, as the arithmetic average of sound levels in the 500-, 1000-, 2000-, and 4000-Hz octave bands. It is generally agreed that the SIL level of the ambient noise is roughly 8 dB less than the overall dBA level.

Table 2-7 shows SIL levels as they affect telephone communications. Figure 2-4 shows how communication ability is affected by the distance between talker and listener and the ambient noise level in SIL.

It is important to note that the SIL is mainly intended to be used in environments with relatively steady noise levels, where speech is not distorted by reverberation or other effects.

Other rating methods currently used for speech intelligibility assessment are the articulation index (denoted AI) and speech transmission index (denoted STI). These methods are used more for sound system analysis than environmental assessments. Both involve much more complex calculation methods than does the SIL, but provide more information in terms of speech intelligibility. The STI is most appropriate for estimating speech intelligibility in reverberant environments. The rapid speech transmission index (denoted RASTI) is a simplified version of STI and instruments have been developed that measure RASTI directly.

Sleep Interference

Much less conclusive research has been performed to determine the effects of noise on sleep interference than in any of the other areas presented herein. The most recent research in this area was conducted by Finegold et al. (1992) for the U.S. Air Force. Because the SIL deals principally with continuous, steady noise sources and sleep disturbance seems to be more related to

TABLE 2-7   Telephone Communication Ability Correlated with Speech Interference Level

SIL (dB)	Telephone
Under 60	Satisfactory
60–75	Difficult
76–85	Very difficult
Over 85	Impossible

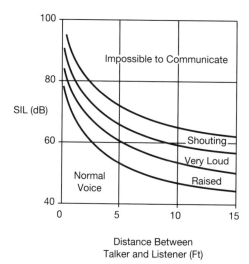

FIGURE 2-4.   Voice levels required for proper communication in different background noise environments.

Distance Between
Talker and Listener (Ft)

startling events that are short in duration with high maximum noise levels, the SEL has been chosen as the most appropriate descriptor in this case. On the basis of their research, as reported in the FICON report (1992), Finegold et ai. have proposed a relationship between percent awakening and indoor SEL. This relationship is shown in Fig. 2-5.

Annoyance

Many studies have been conducted to evaluate numerically the annoyance of noise. Schultz (1978) has produced the most widely accepted annoyance criteria relating to noise from transportation sources such as aircraft, surface vehicles, and trains. The relationship that Schultz established was between the L_{dn} and the percent of people highly annoyed (denoted %HA), based on many earlier studies. Only the most highly annoyed people were used in the analysis because there is the least amount of variability in their responses to the noise. According to Schultz, this minimizes the subjective nature of the responses and provides the best possibility for revealing a clear relationship between noise levels and annoyance. Kryter (1982) disagreed with Schultz on several fronts, most notably that annoyance should be rated in terms of degrees of annoyance instead of only for the most highly annoyed and that different transportation sources produced different annoyance ratings. Schultz's findings still seem to be the most widely accepted among most agencies.

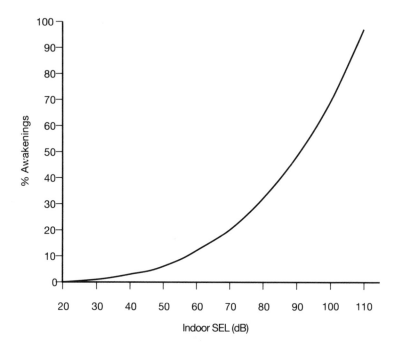

% Awakenings	SEL (dB)
0	20
1	30
3	40
6	50
12	60
20	70
32	80
48	90
69	100
97	110

FIGURE 2-5.   Relationship between levels of sudden noises and sleep interference.   (Data from Finegold et al., 1992; figure adapted from FICON, 1992.)

The so-called Schultz curve has been updated (Fidell et al., 1991) from additional studies and this updated plot is shown in Fig. 2-6. This curve is regarded as the best available prediction of community annoyance in response to transportation sources.

In 1990, the Federal Interagency Committee on Noise (FICON) was formed to review policies related to airport noise. On the committee were representatives from such federal agencies as the Environmental Protection

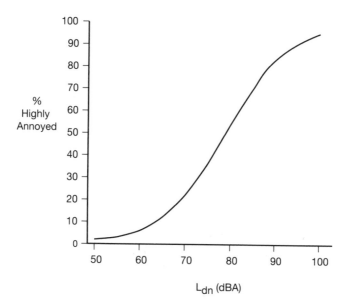

% Highly Annoyed	L_{dn}
2	50
3	55
6	60
12	65
22	70
36	75
54	80
70	85
83	90
95	100

FIGURE 2-6.   Relationship between ambient noise levels and annoyance.   (Adapted from Fidell et al., 1991.)

TABLE 2-8   Effects of Noise on People

L_{dn} (dBA)	Effects		Average Community Reaction	General Community Attitude Towards Area
	Hearing Loss	Annoyance (%HA)		
≥75	May begin to occur	37	Very severe	Noise is likely to be the most important of all adverse aspects of the community environment
70	Will not likely occur	22	Severe	Noise is one of the most important adverse aspects of the community environment
65	Will not occur	12	Significant	Noise is one of the important adverse aspects of the community environment
60	Will not occur	7	Moderate	Noise may be considered an adverse aspect of the community environment
≤55	Will not occur	3	Slight	Noise considered no more important than various other environmental factors

Source: FICON (1992).

Agency, the Federal Aviation Administration, the Department of Defense, the Department of Housing and Urban Development, the Department of Veterans Affairs, the Department of Justice, the Council on Historic Preservation, and the Council on Environmental Quality. The FICON final report, issued in August 1992, includes a consensus summary table of the effects of noise on people in residential areas. This is shown in Table 2-8.

The surveys and analyses above are based principally on sound level limits and therefore must assume specific ambient sound levels. Especially in cases in which ambient sound levels are low (less than 55 dBA), specific increases in ambient noise levels caused by introducing new noise sources must be considered in addition to the absolute resultant ambient level. The nature (steady, intermittent, or impulsive) of the source also must be considered. The annoyance potential of introducing significant new noise sources into an environment is evaluated by environmental assessments and environmental impact statements. For individual noise disturbances, local ordinances have been or can be developed to define the local annoyance criteria. Environmental assessments, noise criteria, and noise regulations and ordinances are discussed in Chapter 5.

REVIEW QUESTIONS

1. The SPL from a siren in an open area is measured to be 90 dBA at a distance of 50 ft.
   a. What is the measured acoustic pressure at 50 ft? at 100 ft?
   b. How far must you move from the siren for it to sound half as loud, assuming no atmospheric or terrain effects, as it does at 50 ft?
   c. At this new location (determined in question 1, b) what SPL would you expect to measure if 50 sirens, identical to the first, were all emitting the same SPL as the original siren simultaneously and from the same location?

2. Four air conditioners are independently emitting SPLs measured to be 65, 70, 68, and 72 dBA at 50 ft.
   a. What would be the total SPL at 50 ft if all four were operating simultaneously?
   b. If, with these four air conditioners operating, a fifth unit were turned on and the resultant total measured SPL at 50 ft were then 79 dBA, what SPL (at 50 ft) is the fifth unit emitting by itself?
   c. Would an average person notice the difference in sound level with the fifth unit operating compared to that with the other four operating?

3. What does it mean if there is a large difference between dBA and dBC readings?

4. Give three examples of invalid measurement quotations; explain why they are invalid.

5. A product is advertised as emitting only half as much noise as its competitor's comparable product. What does this really mean and why is this confusing?

6. A product is advertised as being quiet because it emits 45 dB. Is this claim valid? Why?

7. When would a spectrum analysis measurement be more appropriate than an overall SPL measurement?

8. What is the difference between ambient and background noise levels?

9. What noise descriptor is accepted by the most agencies for use in environmental assessments? Why? Why is the SEL usually higher than other rating methods?

10. How would an L_{dn}, SEL, and SIL of 75 (in the appropriate units) rate in terms of annoyance, sleep interference, and speech interference, respectively?

References

Cherow, E. 1991. *Combatting Noise in the '90s: A National Strategy for the United States.* Rockville, MD: American Speech-Language-Hearing Association.

DoD. 1978. *Planning in the Noise Environment.* AFM 19-10. Washington, D.C.: U.S. Department of Defense.

EPA. 1974. *Information on Levels of Environmental Noise Requisite to Protect Public Health and Welfare with an Adequate Margin of Safety.* EPA 550/9-74-004. Washington, D.C.: U.S. Environmental Protection Agency Office of Noise Abatement and Control.

EPA. 1981. *Noise Effects Handbook: A Desk Reference to Health and Welfare Effects of Noise.* EPA 550/9-82-106. Washington, D.C.: U.S. Environmental Protection Agency Office of Noise Abatement and Control.

FICON. 1992. *Federal Agency Review of Selected Airport Noise Analysis Issues.* Washington, D.C.: Federal Interagency Committee on Noise.

Fidell, S., D.S. Barber, and T.J. Schultz. 1991. Updating a dosage–effect relationship for the prevalence of annoyance due to general transportation noise. J. Acoust. Soc. Am. 89(1):221–233.

Finegold, L.S., C.S. Harris, and H.E. von Gierke. 1992. Applied acoustical report: criteria for assessment of noise impacts on people. J. Acoust. Soc. Am. (submitted).

ISO. 1990. *Acoustics—Determination of Occupational Noise Exposure and Estimation of Noise-Induced Hearing Impairment.* Standard 1999, 2nd Edition. Geneva: International Organization for Standardization.

Kryter, K.D. 1982. Community annoyance from aircraft and ground vehicle noise. J. Acoust. Soc. Am. 72(4):1222–1242.

LGL Ecological Research Associates, Inc. 1991. *Effects of Noise on Marine Mammals.* MMS 90-0093. Herndon, VA: U.S. Department of the Interior Minerals Management Service.

Newman, S.J. and K.R. Beattie. 1985. *Aviation Noise Effects.* FAA-EE-85-2. Washington, D.C.: U.S. Department of Transportation Federal Aviation Administration.

Schultz, T.J. 1978. Synthesis of social surveys on noise annoyance. J. Acoust. Soc. Am. 64(2):377–405.

Suter, A. H. 1991. *Noise and Its Effects.* Technical Appendix to Shapiro, S.A. *The Dormant Noise Control Act and Options to Abate Noise Pollution.* Washington, D.C.: Administrative Conference of the United States.

Taylor, W. J. Pearson, A. Mair, and W. Burns. 1965. Study of noise and hearing in jute weaving. J. Acoust. Soc. Am. 38:113–120.

3

Noise Measurement

SOUND MEASUREMENT DEVICES

As is mentioned in Chapter 2, the instruments most commonly used for environmental noise assessment are sound level meters and spectrum analyzers.

Sound Level Meters

The American National Standards Institute (ANSI) has published standards on types of meters and methods of sound measurement. In ANSI Standard S1.4-1983, *American National Standard Specification for Sound Level Meters*, three types of meters are defined: type 0, having the most stringent tolerances (± 0.7 dB between 100 and 4000 Hz), targeted for laboratory use; type 1, called a precision meter (with tolerances of ± 1 dB between 100 and 4000 Hz); and type 2, a general-purpose meter, having the least stringent tolerances acceptable (± 1.5 dB between 100 and 1250 Hz, up to ± 3 dB at 4000 Hz) for sound pressure level (SPL) monitoring. All sound level meters conforming with these standards would have such information printed either on the meter or in the operator's manual. Sound level meters that do not conform with at least type 2 tolerances are usually considered unacceptable for SPL monitoring.

Most sound level meters have three measurement speeds: slow, fast, and impulsive. Slow measurement speed has an average integration time of 1 s/reading and the fast speed has an average integration time of 125 ms (1/8 of a second). The impulsive speed, with an integration time of 35 ms (roughly 1/30 of a second) for the onset or rise time of a signal, is normally used for assessing human loudness response to impulsive (less than 1 s in duration)

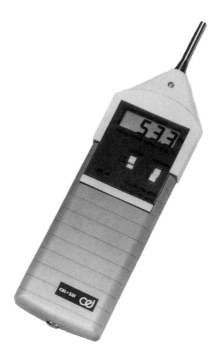

FIGURE 3-1.   Sample basic sound level meter.   (Courtesy of Lucas Industrial Instruments, Severna Park, MD. With permission.)

sounds. The slow speed is usually recommended for environmental noise assessments in which sounds fluctuate by more than ± 3 dB over periods of time and can be averaged with more reliability. The fast speed is usually recommended to monitor discrete events to obtain a clear indication of maximum levels.

Figures 3-1 and 3-2 show typical sound level meters available that meet ANSI standards.

Spectrum Analyzers

When an overall measurement does not provide the required frequency analysis for noise measurements, spectrum analyzers should be used. There are generally two types of spectrum analyzers: octave band and narrow band.

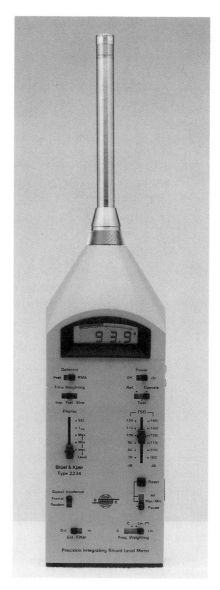

FIGURE 3-2. Sample sound level meter.   (Courtesy of Brüel & Kjaer Instruments, Inc., Marlborough, MA. With permission.)

Octave Band Analyzers

As mentioned in Chapter 2, ANSI Standard S1.6-1984 establishes the common octave band center frequencies used in octave band analyzers. For each octave band, a filter is set up that has a flat (output = input) response over a limited frequency range and a sensitivity that drops off at a rapid rate at the upper and lower frequency limits. These upper and lower frequency limits occur at the frequencies where the response has dropped by 3 dB below the flat response level, known as the 3-dB down point.

The definition of an octave band is that the upper frequency limit of the band (f_u) is twice the lower frequency limit of the band (f_l) and the bandwidth (BW) is the difference between f_u and f_l. As mentioned in Chapter 2, each octave band is identified by a geometric mean frequency of the band, known as the center frequency (f_c). The relationship between f_u, f_l, and f_c, for any bandwidth, is

$$f_c = \sqrt{f_u f_l} = (f_u f_l)^{1/2} \tag{3-1}$$

Figure 3-3 shows typical response curves for commonly used octave band filters. The actual bandwidths are not equal but appear equal in this diagram because they are plotted on a logarithmic scale. Notice from Fig. 3-3 that each filter overlaps with its successor and predecessor at the 3-dB down point of each filter. This shows that f_u for one band is f_l for the following band. The resulting sound measurement, using this system of filters, produces

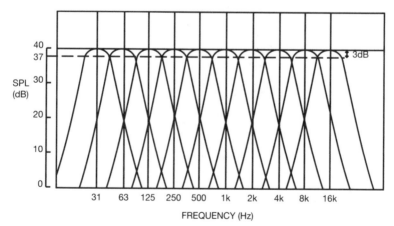

FREQUENCY CHARACTERISTICS OF 10 ADJACENT FULL OCTAVE BAND FILTERS.

FIGURE 3-3.   Frequency response of 10 common octave band filters.

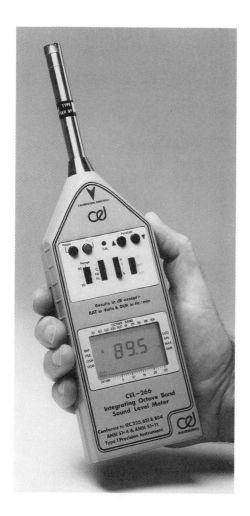

FIGURE 3-4.   Sample octave band analyzer.   (Courtesy of Lucas Industrial Instruments, Severna Park, MD. With permission.)

SPLs for different frequencies in the sound spectrum. This is why this type of analysis is known as spectrum analysis.

Typical octave band analyzers that meet ANSI standards are shown in Figs. 3-4 through 3-6.

Narrow Band Analyzers

Analyzers having frequency bandwidths smaller than those of octave bands are known as narrow band analyzers. If this finer frequency resolution is required in the analysis, 1/3-octave band analysis is available as the next step below octave band analysis. This divides each octave band into geome-

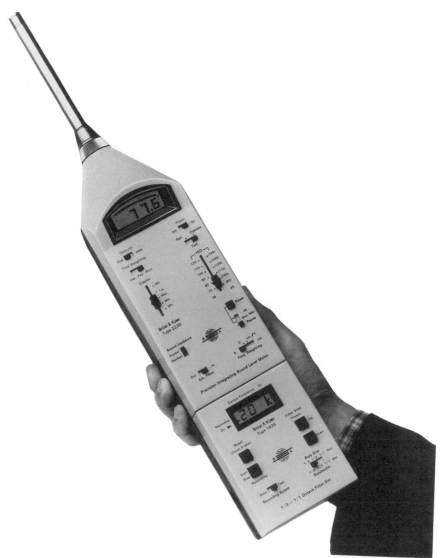

FIGURE 3-5.   Sample octave band analyzer.   (Courtesy of Brüel & Kjaer Instruments, Inc., Marlborough, MA. With permission.)

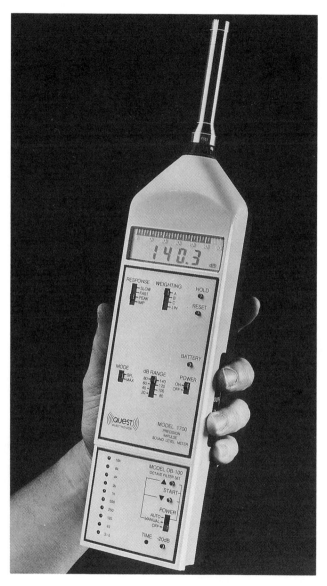

FIGURE 3-6.   Sample octave band analyzer.   (Courtesy of Quest Electronics, Oconomowoc, WI. With permission.)

trical thirds, such that $f_u = 2^{1/3}f_1$, or $f_u = 1.26f_1$. To view spectra in finer frequency resolutions, spectrum analyzers are available with bandwidths down to 1 Hz or less. An example of an instrument that performs narrow band analysis is shown in Fig. 3-7.

When the sound spectrum does not consist of any dominant pure tones (known as a broadband spectrum), any change in frequency bandwidth will cause a change in the SPL in each band, because the amount of sound energy varies with bandwidth. Larger bandwidths cover a larger frequency range and therefore a larger amount of energy for a broadband spectrum. If a source consisted of a pure tone, the energy in the frequency band containing the tone would not change as the bandwidth changes. However, this is not typically the case and the analysis described below applies to the typical broadband spectrum.

If SPL_1 is the SPL of a sound source within a 1-Hz bandwidth and SPL_{band} is the SPL of the source in a larger bandwidth, the following equation can be used to estimate the SPL in different bandwidths:

$$SPL_1 = SPL_{band} - [10 \times \log(BW)] \tag{3-2}$$

For example, say we know that the 1000-Hz octave band SPL of a source

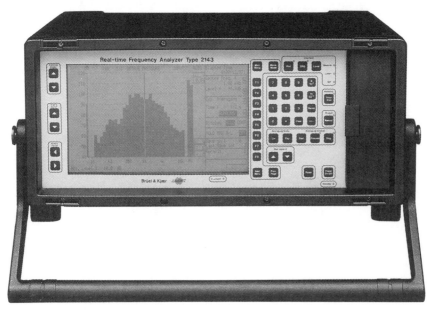

FIGURE 3-7.   Narrow band analyzer.   (Courtesy of Brüel & Kjaer Instruments, Inc., Marlborough, MA. With permission.)

is 70 dB and we want to know what the 1/3-octave SPL would be at the same frequency. The first thing we would need to do is calculate the bandwidth of the 1000-Hz octave band. Using Eq. (3-1) and the octave band definition that $f_u = 2f_1$, Eq. (3-1) would become $f_c = (2f_1f_1)^{1/2} = (2f_1^2)^{1/2} = f_1(2)^{1/2}$. Therefore, for an octave band:

$$f_c = f_1\sqrt{2} \quad \text{or} \quad f_1 = f_c/\sqrt{2} \tag{3-3}$$

Solving Eq. (3-3) for the 1000-Hz octave band, $f_1 = 1000/(2)^{1/2} = 707$ Hz. Because f_u is twice f_1 for an octave band, $f_u = 2 \times 707 = 1414$ Hz. The bandwidth would be the difference between f_u and f_1, $f_u - f_1 = 1414 - 707 = 707$ Hz. Now that we know the bandwidth of the 1000-Hz octave band, we can solve Eq. (3-2) for SPL_1:

$$\text{SPL}_1 = 70 \text{ dB} - [10 \times \log(707)] = 70 \text{ dB} - 28.5 \text{ dB} = 41.5 \text{ dB}$$

Note that arithmetic addition and subtraction are used in these operations, unlike the logarithmic operations discussed in Chapter 2 for combining decibels. Also note that accuracy to tenths of decibels is used for intermediate calculations. Although fractions of decibels are meaningless to hearing perception, rounding off to whole numbers throughout calculations may introduce errors in final answers. Accuracy to a tenth of a decibel should thus be used in these calculations and the final answer can then be rounded to the closest whole number.

Now that we know that $\text{SPL}_1 = 41.5$ dB, we must solve Eq. (3-2) again, this time for SPL_{band}, where the bandwidth would correspond to the 1000-Hz 1/3-octave bandwidth. As stated above, $f_u = 1.26f_1$ for a 1/3-octave band. Using this relationship in Eq. (3-1), we would have $f_c = (1.26f_1f_1)^{1/2} = (1.26f_1^2)^{1/2} = f_1(1.26)^{1/2}$. Therefore, for a 1/3-octave band:

$$f_c = f_1\sqrt{1.26} \quad \text{or} \quad f_1 = f/\sqrt{1.26} \tag{3-4}$$

Solving Eq. (3-4) for the 1000-Hz 1/3-octave band, $f_1 = 1000/(1.26)^{1/2} = 891$ Hz. Because f_u is 1.26 times f_1 for a 1/3-octave band, $f_u = 1.26 \times 891 = 1123$ Hz. The bandwidth would then be the difference between f_u and f_1, $f_u - f_1 = 1123 - 891 = 232$ Hz. Now that we know the bandwidth of the 1000-Hz 1/3-octave band, we can solve Eq. (3-2) for SPL_{band}, using the information just calculated. Rearranging Eq. (3-2) to solve for SPL_{band} we would have

$$\text{SPL}_{\text{band}} = \text{SPL}_1 + [10 \times \log(\text{BW})] = 41.5 \text{ dB} + [10 \times \log(232)]$$
$$= 41.5 \text{ dB} + 23.7 \text{ dB}$$

The 70-dB 1000-Hz octave band level would then be 65 (rounded from 65.2) dB if measured in the 1000-Hz 1/3-octave band. This shows that the SPL in a frequency band would be different for different bandwidths. For this reason, measured and predicted values should all be in terms of the same bandwidth whenever possible. If this is not possible, the additional calculations, such as those performed above, must be performed to ensure compatibility of data.

Listed below, in alphabetical order, are companies that manufacture, service, and sell noise-monitoring equipment that meets ANSI standards.

Brüel & Kjær Instruments, Inc.
2364 Park Central Boulevard
Decatur, GA 30035

Ivie Technologies, Inc.
1366 West Center Street
Orem, UT 84057

Larson Davis Laboratories
1681 West 820 North
Provo, UT 84601-1341

Lucas Industrial Instruments
760 Ritchie Highway, Suite N6
Severna Park, MD 21146

Quest Electronics
510 South Worthington
Oconomowoc, WI 53066

Scantek, Inc.
916 Gist Avenue
Silver Spring, MD 20910

Sound Intensity Systems

In mathematical jargon, sound intensity is a vector quantity whereas sound pressure is a scalar quantity. In plain English this means that sound pressure refers to an amount, or magnitude, of sound energy whereas sound intensity refers not only to the amount of energy, but also to the direction in which it is traveling. Whereas one microphone can be used to measure the scalar sound pressure level, a minimum of two microphones close together, monitoring simultaneously, is required to monitor the vector sound intensity level. A typical intensity probe, the instrument that measures sound intensity, consists of two microphones, arranged either side by side or facing each other, connected to their associated signal processing equipment. This equipment usually consists of preamplifiers hooked into a pistol grip, hooked in turn to a spectrum analyzer.

Sound intensity systems are complicated to use because there is much signal processing and interpreting to be done to produce reliable results. The systems are also expensive, beginning at around $10,000 for the least expensive one. The systems are not simple or user friendly. The operator must be fully trained in the technology and operation of the system to provide meaningful results. When performed properly, sound intensity measurements can provide valuable insight into source location when it is not easily discernible; however, experienced users should be the only ones performing and interpreting measurements because it is much easier to extract invalid than valid data from this measurement technique unless the operator is knowledgeable about the system.

Sound intensity systems are used most often either to localize noise sources or to measure the sound power of a source. Unless sound source identification is difficult, sound intensity systems are rarely used in environmental noise assessments.

SOUND MEASUREMENT TECHNIQUES

Before discussing noise measurement methods, a brief description of sound fields would be helpful.

Sound Fields

A sound field refers to the acoustic conditions in the measuring environment. The near field is the region close to a sound source (usually considered within one-quarter of the largest wavelength of interest), where levels fluctuate drastically (up to and exceeding ± 10 dBA) with small changes in distance from the source. Noise levels for environmental assessments should not be taken within the near field of a source for this reason. A type of near field measurement would be within one-quarter of a wavelength of the lowest frequency (the largest wavelength) of a reflective wall or surface. In this case, a microphone would be measuring two pressure waves at the same time, one incident on the surface and the other reflecting off the surface. Therefore, twice the acoustic pressure, translating to an additional 6 dB of SPL, would be monitored close to the surface (recall from Chapter 2, that doubling the sound pressure causes a 6-dB SPL increase while doubling the sound power, or doubling the number of incoherent sources, causes a 3-dB increase in SPL). This artificial amplification of the source makes taking measurements close to reflecting surfaces not recommended in practice, unless specified.

As we move farther from a point source than the near field limits, we would enter a region where the SPL drops off at the inverse square law rate of 6 dB per doubling of distance from the source. To calculate the SPL at any distance away from a point source under these conditions, the inverse square law states that

$$SPL_1 - [20 \times \log(d_2/d_1)] = SPL_2 \qquad (3\text{-}5)$$

where SPL_1 is the sound pressure level at the location closer to the sound source, SPL_2 is the sound pressure level at the location farther from the source, d_2 is the distance from the source where SPL_2 was measured, and d_1 is the distance from the source where SPL_1 was measured.

This region is known by one of three names, depending on the application. The name *far field* contrasts this region from the near field, and is normally used to specify that measurements were not taken in the near field. This region can also be called the *direct field* to contrast it from the reverberant or diffuse fields defined below. Direct field measurements imply that readings are indicative of those generated only by the source of interest, without the contributions of reflections that would add to the measured level. These types of measurements are necessary when noise levels of equipment are measured for rating purposes within a room. The third name for this type of environment is the *free field*, where there are no obstructing surfaces in the sound path of spherical wave propagation. Free field conditions usually exist in large open spaces outdoors or in rooms with walls that are acoustically highly absorptive. Near, far, and free fields are commonly referenced for both indoor and outdoor environments. All other fields mentioned herein are normally referenced for indoor environments.

A room designed for free field measurements is known as an anechoic (without echoes) chamber. This type of room has highly absorptive material covering its walls, so that nearly all of the acoustic energy striking the walls is absorbed. In this way, no significant reflections exist in the room and the sound emitted directly from the source of interest alone can be measured. The ideal anechoic chamber would absorb all incident sound energy at its walls, so that no sound is reflected off the walls. Although this is not physically possible because some sound always reflects off an obstacle, the walls of anechoic chambers usually absorb enough sound energy to produce the kind of environment intended for measurements. Anechoic chambers have ideal environments designed especially for laboratory-type measurements and would not be found in typical buildings. An example of an anechoic chamber is shown in Fig. 3-8. The absorptive material on the walls

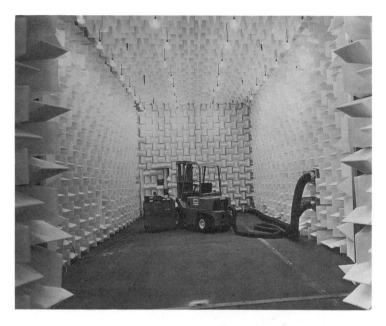

FIGURE 3-8.   An anechoic chamber. Note that a true anechoic chamber would also have absorptive wedges on the floor, with acoustically transparent wire mesh over it to support people or testing equipment.   (Courtesy of Industrial Acoustics Company, Inc., Bronx, NY. With permission.)

of the anechoic chamber is formed into wedge shapes to trap any reflected sound and provide the maximum absorption possible.

On the other end of the scale of ideal room environments for measuring sound are reverberation chambers. A reverberation chamber would have most of its space in a reverberant or diffuse field. In a diffuse field, there are so many reflections contributing to the total sound field that the sound level measured would be the same anywhere within the room. For this type of acoustic environment to exist, all of the walls of the room would have to be highly reflective. Reverberation chambers are typically used to measure a sound source level on one side of a partition being tested for its sound insulation properties, or in any other use where a sound level is needed to be measured independent of location. The ideal reverberation chamber would reflect all sound energy off its walls so that no sound is absorbed and the sound energy remains constant within the room. Although this is not physically possible, because some sound is always absorbed when it is redirected off a partition, the walls of reverberation chambers usually reflect

FIGURE 3-9.   A reverberation chamber.   (Courtesy of Industrial Acoustics Company, Inc., Bronx, NY. With permission.)

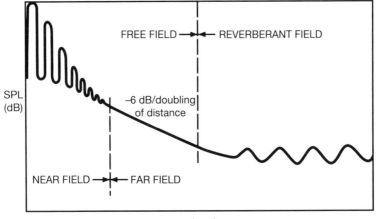

FIGURE 3-10.   Sound fields in rooms relative to distances from a source.

enough sound energy to produce the kind of environment intended for measurements. As for anechoic chambers, reverberation chambers have ideal environments designed especially for laboratory-type measurements and would not be found in typical buildings. An example of a reverberation room is shown in Fig. 3-9.

Rooms in actual buildings are compromises between anechoic and reverberation chambers. Such a combination of sound fields is illustrated in Fig. 3-10. Rooms that have larger diffuse fields than free fields are usually categorized as being "live" and those having much larger free fields than diffuse fields are usually classified as being "dead". Life, then, in terms of room acoustics terminology, is classified by the amount of reflectiveness, or the amount of sound bouncing around the room.

Effects of Close Reflective Surfaces

When a sound source is hanging in free space with no reflective surfaces or obstructions in its vicinity, sound waves spread out spherically from the source, as discussed in Chapter 1. When a large reflective surface is near the source, it still emits sound in a spherical pattern but its level is amplified because of the reflected sound. Large reflecting surfaces also change the directivity, or directional propagation characteristics, of the sound source. These principles are used in loudspeaker horn design, as is discussed in Chapter 6.

A sound source that emits equal amounts of sound in all directions in a spherical pattern is called an omnidirectional source. When an omnidirectional sound source is on a reflective floor or wall, the source level would be doubled above the floor or near the wall because the half of the spherical sound wave that would have traveled in the direction of the reflective surface is reflected back in the opposite direction to add to the half of the spherical wave already traveling in that direction.

Along the same lines, if an omnidirectional source were placed at a room edge where two large reflective surfaces meet, the spherical wave would be confined to one-quarter of its normal radiating area and the resulting level would be multiplied by a factor of four. Going one step further, moving an omnidirectional source into a corner where three large reflective surfaces meet, the spherical wave would be confined to one-eighth of its normal radiating area and the resulting level would be multiplied by a factor of eight.

The directivity factor, normally denoted by the letter Q, is the ratio of sound intensities of the spherical wave in its area reduced by large reflective surfaces to that in open space. Q can be calculated from Eq. (3-6):

$$Q = 2^n \tag{3-6}$$

where n is the number of large reflecting surfaces near the omnidirectional source. The directivity index, D, is the SPL that would be added to that occurring in free space when an omnidirectional source is placed near large reflective surfaces. The directivity index is:

$$D = 10 \times \log(Q) \tag{3-7}$$

Therefore, 3 dB would be added when an omnidirectional source is placed on a reflective floor, 6 dB would be added when it is placed at a room edge, and 9 dB would be added when it is placed in a corner. For an omnidirectional source in free space, Q is 1 and D is 0 dB.

The SPL and sound power level (L_W) are not interchangeable quantities because the SPL depends on distance from the source and L_W does not. When an omnidirectional source is in a free field and near a reflective surface, the relationship between SPL and L_W is as follows:

$$\text{SPL} = L_W - [20 \times \log(r)] + D - 1 \tag{3-8}$$

where r is the distance, in feet, between the source and SPL monitoring location. Therefore, for example, if a sound source were outdoors and close to the ground the SPL could be approximated by $L_W - [20 \times \log(r)] + 2$. When a source is in a typical room, these approximations must be supplemented by methods that compensate for reflections and the absorption characteristics of the terminating surfaces of the room.

Standard Methods

Acoustic measurement, rating, and testing methods are standardized by the American National Standards Institute (ANSI) and the American Society for Testing and Materials (ASTM). ANSI and ASTM standards are written by committees of volunteer acoustic professionals and are intended to be updated or reaffirmed every 5 years. All standards are extensively reviewed and are not issued until approved by a consensus of experts. The standards provide guidelines that are legally binding only when referenced in regulations or other legally binding documents. Referencing compliance with these standards applies credibility to data acquired in studies.

The most applicable ANSI standards for the types of noise measurements used in environmental noise assessments are ANSI S1.13-1971(R1986), *American National Standard Methods for the Measurement of Sound Pressure Levels*, and ANSI S12.9-1988, *American National Standard Quantities and Procedures for Description and Measurement of Environmental Sound*, Part 1. The most relevant ASTM standard for environmental noise measurements is ASTM E 1014-84(1990), *Standard Guide for Measurement of Outdoor A-Weighted Sound Levels*. In addition, the Society of Automotive Engineers, Inc. (SAE), published Aerospace Recommended Practice (ARP) 866A, *Standard Values of Atmospheric Absorption as a Function of Temperature and Humidity*, in 1975. This provides atmospheric absorption information that must be applied to sound levels at distances away from sources most typically used for predicting aircraft noise effects. The SAE also provides standards for monitoring in vehicular noise emission certification. ANSI S1.26-1978 (R1989), *American National Standard Method for the Calculation of the Absorption of Sound by the Atmosphere*, provides information similar to SAE ARP 866A.

These standards present guidelines for SPL measurement, calculation, and reporting practices to provide reliable data. Note that the last digits in the ANSI and ASTM standard designations correspond to the year in which the standard was written or last reaffirmed (R standing for reaffirmed). The most current version of each standard should be used for each application. The standards referenced herein are the latest versions as of this writing.

Information on these and other acoustic standards can be obtained from the following locations.

ASA (Acoustical Society of America) Standards Secretariat
335 East 45th Street
New York, NY 10017-3483

ANSI
11 West 42nd Street, 13th Floor
New York, NY 10036

ASTM
1916 Race Street
Philadelphia, PA 19103-1187

SAE
400 Commonwealth Drive
Warrendale, PA 15096-0001

Information unique to each standard is mentioned below. All common recommendations are listed separately in the section Industry-Accepted Measurement Practices (below).

ANSI S1.13-1971(R1986)
ANSI S1.13-1971(R1986) divides measurement methods into three categories: survey, field, and laboratory. The survey method is the least accurate,

using a hand-held sound level meter (meeting at least type 2 requirements) and recording overall sound pressure levels in the given environment. If the sound level fluctuates by more than ± 3 dB, using the slow meter speed, the maximum and minimum readings should be recorded to provide the range of data. If the difference between the SPL with the source in question operating and the background SPL is between 4 and 15 dB in any frequency band, Table 3-1 should be used to derive the actual source level with the contributions of all other sources subtracted. If this difference is less than 4 dB, the standard states that the source level cannot be derived accurately enough from the comparable levels of all other sources.

Although the corrections listed in Table 3-1 go down to 0.1 dB, it is seldom necessary to use corrections of less than 0.5 dB for these types of measurements. The survey method is appropriate for most environmental noise assessments.

The field method specifies the use of a type 1 octave or narrow band analyzer to provide spectrum analysis of the noise in question either without changing the source environment or with minor changes to the source environment to reduce the effects of sound path obstructions external to the source in question. The measuring microphone should be mounted on a tripod or suspended in some manner, with the observers and associated equipment located outside the test area. No weighting should be used in the measurements and the range of minimum to maximum level should be recorded when noise levels fluctuate over a wide range. Table 3-1 can be used for background noise subtraction, as for the survey method.

TABLE 3-1   Standard Corrections for Ambient Sound Pressure Levels

Difference (in dB) between SPL Measured with Sound Source Operating and Background SPL	Correction (in dB) to be Subtracted from SPL Measured with Sound Source Operating to Obtain SPL from Sound Source Alone
4	2.2
5	1.7
6	1.3
7	1.0
8	0.8
9	0.6
10	0.4
11	0.3
12	0.3
13	0.2
14	0.2
15	0.1

Source: ANSI S1.13-1971(R1986). (From the American National Standards Institute, New York, NY. Reprinted with permission.)

The laboratory method requires the use of type 0 instrumentation for measurements in indoor controlled environments such as anechoic and reverberation chambers. This method is used primarily for specific source measurements for laboratory rating. Most other measurement requirements for the laboratory method are similar to those specified for the field method.

In addition to these methods, the standard discusses measurement practices for steady (constant in level) and nonsteady (fluctuating, intermittent, or impulsive in nature) sources. Impulsive noises are defined as those lasting less than 1 s. Also discussed are environmental factors that must be taken into account for any noise measurements. The principal environmental factors that are required to be observed are distances from obstacles or reflecting surfaces, microphone locations, and atmospheric and terrain conditions outdoors. In terms of obstacles or reflecting surfaces, none having dimensions greater than one-quarter of a wavelength (except for the ground) of the lowest frequency of interest should be within five wavelengths (of that same lowest frequency) of the source. In addition, these same obstacles should not be within five wavelengths of the monitoring microphone or five times the distance between the source and microphone, whichever is greater. This is illustrated in Fig. 3-11. For example, if 250 Hz (having a wavelength of approximately 4 ft) is the lowest frequency of interest, any obstacles having

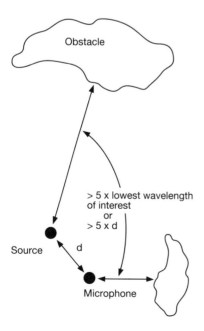

Obstacle

> 5 x lowest wavelength
of interest
or
> 5 x d

Source

d

Microphone

FIGURE 3-11.   Recommended locations of sound sources and measuring microphones relative to obstacles, according to ANSI S1.13-1971(R1986).

dimensions larger than 1 ft should be no closer than 20 ft from the source or microphone as long as the monitoring microphone is within 4 ft of the source. If the measurement is 25 ft from the source, no obstacles should be within 125 ft of the monitor.

Preferred microphone positions are 5 ft off the ground or floor. Other heights are acceptable, depending on the situation. For example, if the noise of interest is at the ear of a seated person, the appropriate measurement height would be approximately 4 ft off the ground or floor. For people lying down (in bedrooms), the microphones should be approximately 2 ft off the ground or floor. When performing measurements by an open window, it is recommended that the microphone be centered on the open window at a distance of approximately 2 ft from the window.

When measuring the sound from a specific source, several microphone layouts are suggested, depending on the source characteristics. These layouts are illustrated in Fig. 3-12.

For outdoor measurements, atmospheric conditions and terrain conditions can have a significant effect on measured sound levels. The standard states that measurements shall take place only where the atmosphere is homogeneous (having no density changes) to a height of 33 ft above the ground or to the height of the source in question, whichever is greater,

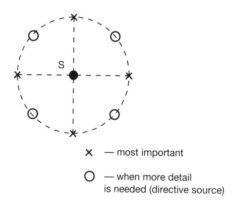

X — most important

O — when more detail
is needed (directive source)

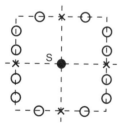

FIGURE 3-12.   Recommended measuring microphone locations with respect to a sound source (S), according to ANSI S1.13-1971(R1986). A directive source would require more locations because it does not radiate sound energy equally in all directions.   (From the American National Standards Institute, New York, NY. With permission.)

temperature changes in only one direction (positive or negative) with height, and wind speeds do not vary with height. Measurements should not be performed when wind speeds exceed 15 miles per hour (mph). Meteorological data to be included in reports should include temperature, relative humidity, barometric pressure, and wind direction and average speed. Terrain conditions must be noted with reported data. All other pertinent information to be noted is mentioned in Industry-Accepted Measurement Practices (below).

ANSI S12.9-1988

ANSI S12.9-1988 defines noise descriptors recommended for use in environmental assessments. These descriptors are defined in Chapter 2. In addition to these definitions, some measurement procedure recommendations not included in ANSI S1.13-1971(R1976) are listed. If the difference between dBA and dBC values exceeds 15 dB, high-level infrasound may be present that can distort readings. This should be noted if it is the case.

Outdoor, free field measurements should not be performed within 50 ft of large reflecting surfaces. If this cannot be avoided, the surface locations should be noted. When outdoor measurements need to be taken near buildings or reflective surfaces, the preferred measurement location is 4 to 6 ft from the facade, 4 ft above each floor level of interest, and, if applicable, centered on the largest window. Preferred microphone locations for indoor measurements are at least 3 ft from any wall, ceiling, floor, or reflecting surface, and at least 5 ft from windows.

ASTM E 1014-84(1990)

ASTM E 1014-84(1990) deals with basic outdoor noise surveys and provides more appropriate guidelines for nontechnical people than the ANSI standards. The ANSI standards are more geared for detailed analysis of complex problems than is this standard. The guidelines presented in this standard are most appropriate for use as general industry-accepted field practices.

ASTM E 1014-84(1990) specifies the use of an ANSI S1.4-1983 type 2 sound level meter, preferably with a capability for headphones to monitor the output of the sound level meter. With headphones, the inspector can hear whether the meter is picking up contaminating signals such as wind noise, electromagnetic radiation from transmission lines, or radio signals. The ASTM standard mentions that acoustic specialists should be consulted if measurements need to be taken more than 1600 ft away from a source to account properly for atmospheric propagation. Natural seasonal sounds, such as those generated by insects or birds, should be noted and measurements should be taken again at a time when these sources do not contribute to the background levels. Measurements should be taken during peak and nonpeak periods of activities. It is important to note that measurements should be taken, whenever possible, both with and without the source in question operating to provide the most accurate account of the contribution of the source in question.

In measuring noise levels, it is recommended that visual averaging should be avoided and instantaneous levels should be recorded. The number of sound levels recorded should be at least 10 times the range between sound levels observed. For example, if a ± 3-dBA range is noted in readings, at least 60 readings should be recorded for an adequate sampling.

No measurement locations should be closer than 5 ft to each other and, if the monitored calibration signal has drifted by more than 1 dB over the course of a monitoring session, the recorded data are not considered valid. Other important points mentioned in this standard are also mentioned in other standards and are listed under Industry-Accepted Measurement Practices (below).

SAE ARP 866A

The ARP 866A standard provides sound attenuation information that compensates for classical and molecular absorption of sound energy in the atmosphere. These effects cause energy dissipation exceeding that provided by the spherical spreading, refraction, and ground cover. These effects vary with temperature and humidity. Their values are listed in ARP 866A, for common octave and 1/3-octave bands, for air temperatures ranging from 1 to 100°F (in 1°F intervals) and for relative humidities ranging from 1 to 100% (in 1% intervals). The attenuation values are listed in terms of decibels per 1000 ft. Table 3-2 shows these values, in terms of their so-called absorption coefficients, for 59°F (the international day temperature at sea

TABLE 3-2   Atmospheric Absorption at 59°F and 70% Relative Humidity

Octave Band Center Frequency (Hz)	Atmospheric Absorption Coefficient (dB/1000 ft)
63	0.1
125	0.2
250	0.4
500	0.7
1000	1.5
2000	3.0
4000	7.6
8000	13.7

Source: SAE ARP 866A. (Reprinted with permission from SAE 866A FEB91 © 1991 Society of Automotive Engineers, Inc.)

level reference condition) and 70% relative humidity (the average relative humidity at major worldwide airports).

ANSI S1.26-1978(R1989)

ANSI S1.26-1978(R1989) provides information similar to that of SAE ARP 866A. SAE ARP 866A is used extensively in the aircraft industry whereas the ANSI standard is recommended for use with any noise sources. Differences between values listed in SAE ARP 866A and ANSI S1.26-1978(R1989) are only for very high frequencies (above 4000 Hz). Therefore, except for the 8000-Hz value (which is slightly higher in ANSI S1.26), Table 3-2 can be used under standard atmospheric conditions to estimate the atmospheric absorption for either standard. SAE ARP 866A also provides a simpler format for finding the specific atmospheric absorption coefficient than does the ANSI standard.

A consistent trend does not exist between atmospheric absorption and temperature or humidity. Therefore, these values should be calculated or derived from tables provided in these standards for nonstandard atmospheric conditions.

Industry-Accepted Measurement Practices

Many measurement methods prescribed in standards are for laboratory or critical measurement venues. When dealing with a typical noise assessment, it is seldom necessary to spend the time or money on the extensive measurement procedures outlined in the standards. The following practices are considered industry accepted for the measurement of noise levels. These are practices that most standards and agencies agree on and can be considered minimum requirements for valid acoustic measurement practices and data reporting.

Basic measurement procedures that these standards and accepted industry practices agree on are windscreen usage, calibration, careful selection of microphone placement, and consideration of meteorological conditions.

When measurements are performed outdoors or in areas where air flow can be sensed, a windscreen designed to fit the specific instrument should be used. The most commonly used windscreens resemble black sponge balls. They have been designed to block wind noise without attenuating the signal being measured. Without a windscreen, air flow over the microphone may artificially raise the signal level sensed by 20 dBA or more and produce erroneous readings. Even with a windscreen in place, wind speeds above 15 mph can cause erroneous readings. Therefore, wind speed should be monitored in breezy areas and readings should not be taken either when wind speeds exceed 15 mph or when instantaneous readings are viewed to

FIGURE 3-13.   Sound level calibrator.   (Courtesy of Quest Electronics, Oconomowoc, WI. With permission.)

fluctuate by more than 5 dBA with no apparent source causing the fluctuation.

It is generally accepted practice that the measuring instrument is calibrated with a calibrator designed to fit the meter and that calibration is performed before and after each series of readings. Typical sound level calibrators are palm-sized cylinders with adapters to fit the measuring microphone of the meter to be calibrated. Figure 3-13 shows a typical calibrator. They generate a pure tone (single frequency) or set of pure tones at a frequency (or set of frequencies) and SPL (or set of SPLs) set at the facilities of the manufacturer. These frequencies and levels should be printed directly on the calibrator. When the sound level meter microphone is fitted properly in the calibrator adaptor and the calibrator is turned on, the meter should read the known level at the known frequency. If the meter reads overall levels, a 1000-Hz calibrator is ideal because A- and C-weighting networks read the same level as the tone at 1000 Hz. If the known level is not indicated on the meter when the calibrator is attached and operating, the meter must be manually adjusted to conform with the calibrator output. With a properly operating meter, manual adjustment should be required infrequently (especially not during a measurement session) and the adjustment should be less than 2 dBA. If the meter consistently drifts out of calibration, it should be returned to the factory for repair.

Bear in mind that corrections would have to be made to the monitored calibration signal level if an A- or C-weighted instrument is calibrated at tones other than 1000 Hz, in accordance with the chart in Table 3-3. The A- and C-weighting response curves are shown in Fig. 3-14.

These corrections can be compared with the average human hearing frequency sensitivity curve shown in Fig. 3-15. Figure 3-15 shows the amplified frequency range of 2000 to 4000 Hz as a bulge in the curve. It also shows that, even though our hearing mechanisms are purported to be sensitive to frequencies down to 20 Hz, a sound at 20 Hz would have to be approximately 70 dB louder than a sound at 1000 Hz for each to appear to have the same loudness.

Calibrators and meters should also be factory calibrated by the manufacturer once each year to ensure that accurate readings are being performed. The manufacturer always provides calibration certificates and usually stamps the instrument with the date of certification and the date recommended for recertification. This information is usually provided with data reports to ensure reliability of reported results. It is also common practice to list the model and serial numbers of all equipment used in the study to keep the possibility open of consistently repeating measurements in the case that questions arise.

TABLE 3-3 A- and C-Weighting Corrections at Octave Band Center Frequencies

Octave Band Center Frequency (Hz)	A-Weighting Correction (dB)	C-Weighting Correction (dB)
63	− 26.2	− 0.8
125	− 16.1	− 0.2
250	− 8.6	0
500	− 3.2	0
1000	0	0
2000	+ 1.2	− 0.2
4000	+ 1.0	− 0.8
8000	− 1.1	− 3.0

Source: ANSI Standard S1.4-1983, Specification for Sound Level Meters. (Reprinted by permission of the Acoustical Society of America, New York, NY.)

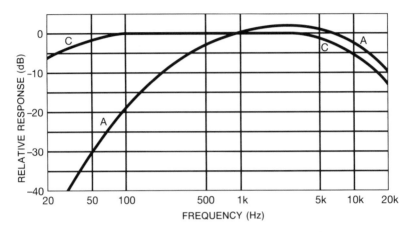

FIGURE 3-14.   The frequency response of A- and C-weighting networks.

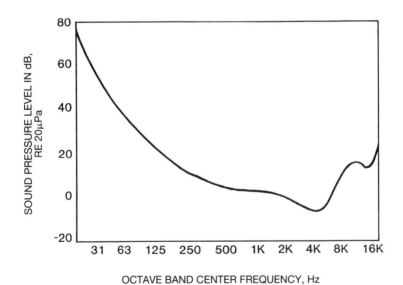

FIGURE 3-15.   Average human hearing frequency sensitivity curve.

When performing noise measurements, it is accepted practice to place the measuring microphone a minimum of 3 to 4 ft away from any reflecting surfaces, including the ground, walls, and the body of the person performing the measurements, to avoid reflections that can contaminate readings. The reason for this is that acoustic pressure doubles close to a reflective surface, adding up to 6 dB to the reading. It is usually preferred, whenever feasible, to mount the meter on a tripod to avoid signal contamination caused by the monitoring personnel.

According to ANSI Standard S1.13-1971(R1986) and accepted industry practice, whenever the SPL with a source in question operating is within 10 dBA of the SPL measured without the source in question operating, a correction must be applied to the total SPL measured to remove the contribution of the background noise and provide the SPL generated by the source in question alone. Table 3-1 lists the corrections that are applied. When the difference between SPLs with the source operating and without it operating is 3 dBA or less, it is usually considered that the SPL of the source in question is equal to or less than the background SPL and the source SPL cannot be accurately derived from the readings.

Table 3-1 represents a refinement of Table 2-3 (which should be used only in general cases), cutting off differences at 4 dBA and listing correction factors to an accuracy of tenths of a decibel as opposed to rounding to whole numbers. The dBA values in terms of tenths of decibels are useful only when calibrating instruments or statistically averaging large volumes of data, because the smallest increment over which the human hearing mechanism recognizes change is 2 or 3 dBA. Changes in terms of tenths of dBA are not discernible to the hearing mechanism. It is appropriate to use tenths of dBA units for L_{dn} and other statistical ratings because they involve large volumes of data recorded over extended periods of time. However, instantaneous data should be reported only in terms of whole decibel values because, in addition to the reasons mentioned above, the accuracy of the monitoring instruments is typically ± 1 dB for the best (type 1) required meters.

In addition to the wind considerations mentioned above, other meteorological conditions are typically noted in measurement reports. These include factors that could affect sound readings, such as general atmospheric conditions, relative humidity, temperature, and barometric pressure. Local weather report records usually provide enough information in these areas. Measurements should not be performed if there is precipitation falling or on the ground during the measurement period, because the moisture may damage the meter and the nature of the noise, especially when surface transportation related, would change from that normally associated with dry thoroughfares.

The tolerances of the instruments should also be investigated from the manufacturer specification sheets to ensure that the measurement conditions (such as relative humidity, temperature, and barometric pressure extremes) are within the tolerances of the instrument.

Accompanying the report of measurement results should be a description and sketch of the measurement site, preferably to scale, including source and measurement locations. If a scale drawing is impractical, all key distances and dimensions must be specified.

For reference, Fig. 5-10 (see Chapter 5) provides a sample blank data sheet that includes all data normally required for field noise studies used in environmental noise assessments.

Calculating A-Weighted Levels from Octave Band Spectra

The correction values in Table 3-3 for A-weighting, rounded to the closest whole number and combined with Table 2-1 (see Chapter 2) for decibel addition, can be used to calculate overall dBA values from unweighted octave band data. Each octave band level is corrected by its associated A-weighted correction factor (in arithmetic mathematics) and the resultant values are combined logarithmically, using Table 2-1 values, to produce the single overall dBA level. An example of this calculation is shown below.

Let us assume an unweighted spectrum is provided to us as follows:

Octave Band Center Frequency (Hz)	Given Spectrum (dB)
63	76
125	72
250	68
500	65
1000	67
2000	62
4000	60
8000	55

First we round off the A-weighted correction values in Table 3-3 and apply them to the given spectrum to derive the A-weighted octave band spectrum. The A-weighted levels for the different octave bands are then combined two at a time, using the guidelines of Table 2-1, until a single number is derived. The specific method of combination does not matter. The resultant dBA value should be the same for any method, within ± 1 dB, as long as each decibel value is used once in the addition process. A full calculation for this example is as follows:

	Frequency (Hz)							
	63	125	250	500	1000	2000	4000	8000
SPL (dB)	76	72	68	65	67	62	60	55
A-weighting correction	−26	−16	−9	−3	0	+1	+1	−1
A-weighted spectrum	50	56	59	62	67	63	61	54
First-level addition		57		64		69		62
Second-level addition			65				70	
Third-level addition					71			

The resultant overall level is then 71 dBA for this example. The C-weighted value can be calculated in the same manner, using the appropriate values from Table 3-3. In that case, the resultant overall level would then be 79 dBC.

REVIEW QUESTIONS

1. A sound source has a 500-Hz octave band SPL of 94 dB measured at a distance of 10 ft.
   a. What is the acoustic pressure at 20 ft, assuming free field conditions?
   b. What is the SPL of this 500-Hz source at 50 ft, measured in a 10-Hz bandwidth, assuming free field conditions?
2. The unweighted octave band spectrum of a source has SPLs of 90 dB in each octave band from 63 to 8000 Hz. Calculate the overall unweighted, dBA, and dBC levels for this source.
3. List 10 considerations that should be included in any noise measurements.
4. The sound pressure level of a sound source is measured to be 95 dB 100 ft away from the source in each octave band. Assuming effects only from atmospheric attenuation and free field conditions, what would be the dBA value at a distance of 2000 ft away from the source?
5. Name the different "types" of sound level meters that are usually acceptable for noise monitoring. What other kinds of instruments are there for measuring noise and in what situations would each be used?

6. Calculate the octave band frequency limits for the eight commonly used bands and plot their frequency responses on a graph of sound level sensitivity versus frequency (on a logarithmic scale).
7. What are the best places to monitor sound levels indoors? outdoors?
8. List five situations in which noise readings should not be taken.
9. What are the different common integration speeds of sound level meters and under what situations would each be used?
10. An operating compressor with a sound power level of 100 dB is sitting on the ground 50 ft from the closest residential property. A local noise ordinance sets a sound pressure level limit of 60 dB at any residential property lines from compressors. Under free field conditions, would this compressor comply with the ordinance? If not and all parties are agreeable, what specification could a barrier have that would provide enough attenuation to allow the compressor to comply with the ordinance?

4

Noise Control Terminology and Design

THE SOURCE–PATH–RECEIVER SYSTEM

In dealing with any situation involving sound perception and control, it is best to consider the process in terms of three parts: source, path, and receiver. The source is the equipment or process directly responsible for the sound generation. The path encompasses all media (such as air, water, or solid materials) that sound waves encounter and react with as they travel from the sound source to the ultimate sound receiver. The receiver is the final destination of concern for the sound in question, which interprets the sound according to its own standards. In the field of environmental noise assessment, the receiver is usually a person being affected. The hearing mechanism of an affected person is actually the final destination of the noise source of concern.

Each of the three independent links of the sound chain mentioned above interacts with the others to produce a final result, as is illustrated schematically in Fig. 4-1. In addition to the path of sound propagation affecting the sound reaching the receiver, the shape and location of the receiver can affect the sound path, or the properties of the path can affect the source. For example, a microphone or person in the sound path may cause enough disruption to the sound path to alter the sound field. In critical measurement situations, and especially in laboratory measurements, such factors must be accounted for. This would typically not need consideration for most environmental noise assessments but should be kept in mind so that only required personnel and equipment are present at the monitoring site. An example of the path affecting the source occurs in ducted ventilating systems and turbomachinery where high sound pressures generated by rotating fans can cause what is known as back reaction, in which the sound pressure in the

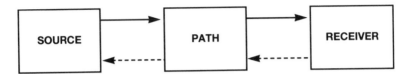

FIGURE 4-1.   The interaction between components of the noise propagation system.

duct generated by the fan(s) reflects back to the fan(s) and changes the pressure, and thus the noise characteristics, of the fan(s).

The control of noise can be accomplished by treating any or all of these parts; however, because of the different physical properties of each part of the noise system, they must be treated separately when noise control is required. It is preferred to control the noise at the source because that would eliminate the noise problem most effectively. If control at the source is not feasible, control in the path should be explored next. Control at the receiver should be used as a last resort because, for the reasons mentioned below, it is the least effective method that often ignores a problem that will eventually have to be solved. Each part of the system is described below, following a discussion of how sound interacts with noise control materials and their associated rating methods.

NOISE CONTROL FUNDAMENTALS

The subject of noise control is discussed in general below. Points are made to give the reader a basic understanding of the noise control options that are available. The discussion only skims the surface of the discipline and the reader interested in more detail or in methods of calculation for specific sources is referred to the references listed at the end of this chapter.

Sound Interaction with a Partition

Basic terminology used in noise control design is explained below. When a sound wave is incident on a barrier, some of the sound energy is absorbed by the barrier material, some is transmitted through the barrier, and some is reflected off the barrier, as is illustrated in Fig. 4-2. High amounts of absorption minimize reflections but do not effectively minimize transmission of sound through the barrier. Therefore, there are two types of noise control designs that account for different circumstances. The distinction is made below between absorption and transmission loss designs. Vibration isolation fundamentals will also be discussed.

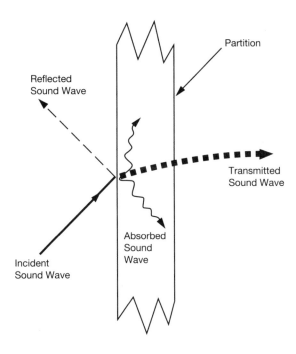

FIGURE 4-2.   Interaction of a sound wave with a partition.

Sound Absorption

Sound absorption occurs when the wavelength of sound waves incident on a partition are small compared to the dimensions of irregular surfaces and the sound energy is dissipated as heat energy as waves bounce around within the partition materials. Porous materials are usually effective sound absorbers. Because absorption is highly dependent on wavelength (size), low frequencies (below 250 Hz) are more difficult to absorb than higher ones and absorption characteristics can change dramatically with frequency. Therefore, whenever a noise problem is judged to require absorption for its solution, the material chosen must have characteristics that are effective in the frequency range of interest.

Applications

Sound absorption is effective in controlling reflections off surfaces within a room. By itself, absorption is not effective in controlling sound transmission through partitions. This is why a separate section follows this discussion of absorption to deal with transmission control. Absorption is an effective noise control solution for echoes and reverberation. As mentioned in Chapter 1, echoes within a room are caused by discrete reflections off surfaces at least 40 to 50 ft from a listener. Reverberation is caused by successive sound reflections off many room surfaces, and can add up to 15 dBA to the noise level in a room.

Absorption Coefficient

Acoustic absorption is usually defined in terms of an absorption coefficient (usually denoted by the Greek letter alpha, α), defined as the ratio of absorbed to incident sound energy from a single interaction between a sound wave and a certain material. Absorption coefficients range from 0 to 1 and vary with frequency. $\alpha = 0$ means that the material absorbs no sound and reflects all sound energy incident on it. $\alpha = 1$ means that the material absorbs all sound energy incident on it and reflects none. Absorption coefficients of 0 and 1 are ideal values that do not exist in reality because all materials reflect and absorb some sound. Some specification sheets on materials quote absorption coefficients greater than 1. This is caused by the testing methods and can be misleading. Any absorption coefficients listed as greater than 1 should be taken as 1 in any calculations or considerations. In general, materials having absorption coefficients less than 0.15 are considered to be reflective and those having absorption coefficients greater than 0.4 are considered to be absorptive.

Absorption Calculations

The total absorption in a room is defined as

$$A = \alpha_1 S_1 + \alpha_2 S_2 + \cdots + \alpha_n S_n \tag{4-1}$$

where α_1, α_2, and so on are the absorption coefficients of each different surface material in a room and S_1, S_2, and so on are the surface areas corresponding to the surfaces having the same subscript number. The units commonly used for total absorption are known as sabins, named after Wallace Clement Sabine (1868–1919), the American physicist who built the foundation for room acoustical analysis. One sabin is defined as the total absorption provided by a 1-ft^2 piece of material having an absorption coefficient of 1 (totally absorptive).

As is mentioned above, α changes with frequency. For typical panel materials, α usually increases as frequency increases.

One of the principal uses for sound absorption is to reduce the size of the reverberant field within a room. This would not reduce the noise level of the sound source in the room but would reduce the reflected sound, and thus reduce the noise levels at locations away from the source. Reverberant field noise reduction (abbreviated NR) has practical limits of 12 to 15 dBA, with typical cases yielding 6 to 8 dBA of noise reduction within a room from proper use of absorption materials. The amount of reverberant field noise reduction to be expected from a specific installation would be

$$NR = 10 \times \log(A_2/A_1)$$ (4-2)

where A_1 is the total absorption in the room before absorptive treatment is installed, and A_2 is the total absorption in the room after absorptive treatment is installed.

An example of this calculation would be instructive. Suppose we have a cubic room with dimensions $10 \times 10 \times 10$ ft. Originally all walls, the floor, and ceiling are covered with material having an α of 0.05 at 500 Hz. In a cubic room, each of the six surfaces has the same area, 10×10 ft (100 ft^2) in this case. The cubic example was chosen for simplicity. For the typical room that is not cubic, the area of each surface would have to be calculated individually. Using Eq. (4-1) to solve for A_1:

$$A_1 = 0.05 \times 600 = 30 \text{ sabins}$$

If an absorptive material having an α of 0.7 at 500 Hz were applied to cover the entire ceiling of this room, leaving all other surfaces unchanged, A_2 can be calculated by Eq. (4-1) as follows:

$$A_2 = (0.05 \times 500) + (0.7 \times 100) = 95 \text{ sabins}$$

Equation (4-2) can then be used with these A_1 and A_2 values to calculate the reverberant field noise reduction resulting from using the ceiling-mounted material. In this case, the NR would be 5 dB at 500 Hz.

Noise Reduction Coefficient

A single number rating system for absorption coefficients over the human speech frequency range is known as the noise reduction coefficient (denoted NRC). The NRC is defined in American Society for Testing and Materials (ASTM) Standard C423-90a, *Standard Test Method for Sound Absorption and Sound Absorption Coefficients by the Reverberation Room Method*, as the average of absorption coefficients for the octave band frequencies of 250, 500, 1000, and 2000 Hz, rounded to the nearest 0.05. This value is useful only for absorption requirements for human speech or sources with dominant frequency components between 250 and 2000 Hz. If human speech is not the only concern, the objectionable frequency range must be identified and, if it is out of the 250- to 2000-Hz range, absorptive materials designed to be effective in the frequency range of interest must be used rather than the NRC value.

ABSORPTION CHARACTERISTICS	NRC	MATERIALS
	-1.0-	
	-0.9-	•Materials designed specifically for high acoustical absorption
	-0.8-	
Highly Absorptive	-0.7-	
	-0.6-	•Typical suspended porous ceiling tile •Typical audience in upholstered seats
	-0.5-	•Heavy curtain •Grass •Upholstered seats
	-0.4-	•Rough soil •Typical audience in wooden or metal seats
Moderately Absorptive	-0.3-	•Heavy carpet on concrete
	-0.2-	•Unoccupied wooden or metal seats •Light multipurpose carpet •Trees
Reflective	-0.1-	•Light curtain •Glass window, wood paneling •Plaster, gypsum board
	-0-	•Smooth concrete, painted brick, marble, glazed tile, water surface

FIGURE 4-3.   General NRC characteristics for common materials.

Figure 4-3 shows common materials with their associated NRC values. Note that NRC values above 0.4 are considered to be highly absorptive. In most offices and dwellings, materials having NRC values in the 0.4 to 0.6 range are sufficient to provide comfortable environments. Materials having NRC values greater than 0.8 are usually much more expensive and are only necessary in rooms having special requirements such as studios, laboratories, or large lecture halls.

Reverberation Time

Intelligibility of sounds and the subjective quality of music and other sounds indoors is typically rated by the reverberation time (denoted T_{60} or RT_{60}). The RT_{60} is the time, in seconds, that it takes for the sound level in a room to decrease by 60 dB after a sound source has stopped emitting sound. The RT_{60} is dependent on the volume of the room and the absorption within the room. Because absorption changes with frequency, the RT_{60} also varies with frequency, normally having the highest values at the lowest frequencies of interest. Mathematically, the RT_{60} is typically described by the following:

$$RT_{60} = 0.05 \times V/A \qquad \text{(4-3)}$$

where V is the volume of the room, in cubic feet, and A is the total absorption in the room, in sabins.

Figure 4-4 shows a general range of ideal midfrequency RT_{60} values for different room uses. Note that rooms used for speech (e.g., lecture halls or playhouses) require much lower RT_{60} values than those used for music. For speech, the lower RT_{60} provides better speech intelligibility and for music, the higher RT_{60} provides greater opportunity for the music to blend in the room.

Equation (4-3) is known as the Sabine formula and is valid for rooms that have large diffuse fields and have a fairly even distribution of absorptive material. When high absorption ($\alpha > 0.8$), low reverberation times ($RT_{60} < 1.0$ s), or a wide variety of absorption coefficients are planned for a room, other RT_{60} formulas must be used that take into account individual materials with their associated absorption coefficients.

Evaluating Vendor Data

Sound absorption data for commercial products come in many forms. The most useful data are provided in terms of α values over the full range of octave band frequencies. In this form, any typical absorption problem can be solved as long as the frequency content of the noise source is known. Many manufacturers offer absorption data in terms of sabins. This can be useful when choosing materials that will be free standing or hanging from

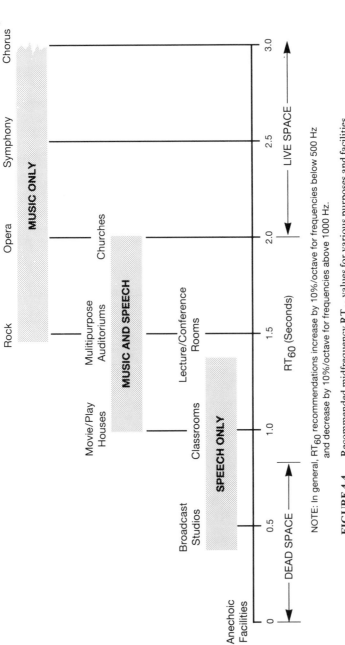

FIGURE 4-4. Recommended midfrequency RT_{60} values for various purposes and facilities.

NOTE: In general, RT_{60} recommendations increase by 10%/octave for frequencies below 500 Hz and decrease by 10%/octave for frequencies above 1000 Hz.

ceilings, because coverage areas are difficult to calculate in these cases. However, these values can be confusing when using them for surface treatment because sabins depend on the surface area available to be treated and the purchaser must work backward to derive the α values to be used in each desired area. Data in this form can save a step in the NR calculation by giving total absorption values; however, it can be misleading. Data given in terms of sabins can mask inflated α values that are listed as greater than 1. This can give certain materials higher absorption ratings than they should have. It must also be considered that, if reflections need to be eliminated through absorption at a specific location, the material must be specified in terms of α, not A, to provide the most effective treatment. As there are no regulations for reporting these data, the consumer must be aware of the absorption coefficients when purchasing equipment that is required explicitly for these purposes.

Table 4-1 lists absorption coefficients and NRC values for common materials.

Sound Transmission

Sound transmission occurs when sound energy passes through a partition. Sound transmission is significantly reduced when the dimensions of the partition are larger than the largest wavelength of incident sound and the partition introduces a sharp change in density into the sound path. The transmission reduction effectiveness of a partition, usually expressed in terms of transmission loss, depends on its mass, flexibility (or stiffness), installation, and design. As for absorption coefficients, transmission loss varies with frequency, usually increasing with increases in frequency. Therefore, whenever a noise problem is judged to require transmission restrictions, the partition chosen must be effective in the frequency range of interest. Other common parameters that deal with transmission reduction are transmission loss noise reduction and insertion loss. These are also dependent on frequency and are described below.

Applications

Design for transmission loss is effective in limiting sound travel between rooms or from one side of a partition to another. This type of design is used for soundproofing, noise enclosures, or acoustic privacy. As was stated above, sound-absorptive materials by themselves are usually not effective in transmission loss. Materials used for high transmission loss can have reflective or absorptive faces, but absorption is usually not their primary purpose.

TABLE 4-1   Acoustic Absorption Performance of Common Materials[a]

	Absorption Coefficient (α)						
	Octave Band Center Frequency (Hz)						
Material	125	250	500	1000	2000	4000	NRC
Brick							
Unglazed	0.03	0.03	0.03	0.04	0.05	0.07	0.05
Unglazed, painted	0.01	0.01	0.02	0.02	0.02	0.03	0.00
Carpet							
1/8-in. pile height	0.05	0.05	0.10	0.20	0.30	0.40	0.15
1/4-in. pile height	0.05	0.10	0.15	0.30	0.50	0.55	0.25
3/16-in. pile and pad	0.05	0.10	0.10	0.30	0.40	0.50	0.25
5/16-in. pile and pad	0.05	0.15	0.13	0.40	0.50	0.60	0.30
Concrete block							
Painted	0.10	0.05	0.06	0.07	0.09	0.08	0.05
Coarse	0.36	0.44	0.31	0.29	0.39	0.25	0.35
Fabrics							
Light velour, 10 oz/yd^2, hung straight at wall	0.03	0.04	0.11	0.17	0.24	0.35	0.15
Medium velour, 14 oz/yd^2, draped to half area	0.07	0.31	0.49	0.75	0.70	0.60	0.55
Heavy velour, 18 oz/yd^2, draped to half area	0.14	0.35	0.55	0.72	0.70	0.65	0.60
Felt, cotton	0.02	0.06	0.17	0.37	0.57	0.63	0.30
Floors							
Concrete or terrazo	0.01	0.01	0.01	0.02	0.02	0.02	0.00
Linoleum, asphalt, rubber, or cork tile on concrete	0.02	0.03	0.03	0.03	0.03	0.02	0.05
Wood	0.15	0.11	0.10	0.07	0.06	0.07	0.10
Wood parquet in asphalt on concrete	0.04	0.04	0.07	0.06	0.06	0.07	0.05
Glass							
1/4-in. sealed, large panes	0.05	0.03	0.02	0.02	0.03	0.02	0.05
24-oz operable windows, closed	0.10	0.05	0.04	0.03	0.03	0.03	0.05
Gypsum board, 1/2-in., nailed to 2 × 4 studs, 16-in. off center, painted	0.10	0.08	0.05	0.03	0.03	0.03	0.05
Marble or glazed tile	0.01	0.01	0.01	0.01	0.02	0.02	0.00
Open plan office panel, 2 in., fiber with fabric, 1.9 lb/ft^2	0.25	0.47	0.71	0.79	0.81	0.78	0.70
Plaster, gypsum, or lime							
Rough finish or lath	0.02	0.03	0.04	0.05	0.04	0.03	0.05
Smooth finish	0.02	0.02	0.03	0.04	0.04	0.03	0.05

(*continued*)

TABLE 4-1   (continued)

Material	\multicolumn{7}{c}{Absorption Coefficient (α)}
	\multicolumn{6}{c}{Octave Band Center Frequency (Hz)}	
	125	250	500	1000	2000	4000	NRC
Hard plywood paneling, 1/4-in. thick, wood frame	0.58	0.22	0.07	0.04	0.03	0.07	0.10
Quilted noise barrier, 3/4 in.	0.05	0.46	0.92	0.83	0.58	0.27	0.70
Resonator block, 8 in., painted	0.20	0.95	0.85	0.49	0.53	0.50	0.70
Spray-on cellulose fiber, 1 in.	0.08	0.29	0.75	0.98	0.93	0.76	0.75
Water surface, no waves	0.01	0.01	0.01	0.01	0.02	0.03	0.00
Wood roof decking, tongue-and-groove, cedar	0.24	0.19	0.14	0.08	0.13	0.10	0.15

[a]Note that the values listed are typical of the general types of materials. For specific applications, it is suggested that one use values associated with the specific materials employed.

Sources: Ceilings & Interior Systems Construction Association (CISCA), 579 W. North Avenue, Suite 301, Elmhurst, IL (1984) (with permission); Hedeen (1980).

Transmission Loss

Transmission loss (denoted TL) is a measure of the amount of sound insulation or reduction provided by a sealed partition between the source and receiver. Just as with absorption, TL is based on a transmission coefficient that varies with frequency. The transmission coefficient (denoted by the Greek letter tau, τ), as the absorption coefficient, ranges from 0 to 1. τ is the ratio of transmitted to incident sound energy resulting from a single interaction between a sound wave and a partition. $\tau = 0$ means that no sound is transmitted through a material, and $\tau = 1$ means that all sound is transmitted through the material or that the material is acoustically transparent. $\tau = 0$ is an ideal limit that would not exist in reality because some sound is always transmitted through a partition. $\tau = 1$, however, can exist. Examples of $\tau = 1$ would be open windows or doors, or openings in walls. In these cases, sound is allowed free passage through the openings.

Transmission loss is defined in units of decibels, by the following equation:

$$TL = 10 \times \log(1/\tau) \tag{4-4}$$

Equation (4-4) shows that TL = 0 when $\tau = 1$, because $\log(1) = 0$. Therefore, a TL of 0 dB implies no sound attenuation. $\tau = 0$ in Eq. (4-4) would yield an infinitely large TL value. The practical upper limit of TL is 70 dB.

As mentioned above, TL generally increases as frequency increases. Over most of the TL spectrum of a partition, TL tends to increase at a rate of 6 dB for each doubling of frequency (octave) and 6 dB for each doubling of partition mass density (mass per unit area). This relationship is known as the mass law. According to the mass law, doubling the thickness of a wall would cause a 6-dB increase in TL. If soundproofing were a major consideration, a 6-dB TL would not be nearly enough for a typical wall, and space and weight considerations usually make this design principle impractical. A more effective method to increase sound insulation between rooms is to use two or more partition walls separated by an air space. The change in density of the media going from air to wall to air to wall to air results in a much greater TL increase than would have been provided by doubling the mass of the partition. Even higher TL values can be achieved when the air space between walls has acoustically absorptive material inside. This material would provide another density change for the sound to travel through and reduce the buildup of sound between walls from internal reflections. Also, where rigid materials would normally be adjacent to one another, a separating air gap or resilient channel breaks the vibration chain through the partition and significantly increases the TL. Such multilayered partitions can provide more than 20 dB more of TL than is achieved by doubling the mass or thickness of a partition in the same amount of space.

This multilayered partition design is used extensively in construction, especially for windows. Double-paned windows, having two window panes with an air gap between them, are commonly used in building construction. Many airplanes use triple-paned windows, having three panes separated by air gaps. When high noise insulation is required, windows are available that have near vacuum conditions between panes. Because sound needs a medium to travel through, a near vacuum (the lack of a medium) would provide significant sound attenuation.

It is important to note that there are frequency response regions in the TL spectrum that do not follow the mass law criteria. Figure 4-5 shows a typical TL spectrum of a partition, showing the response of different characteristic segments of the spectrum. Above and below the mass law region of the TL spectrum are regions affected by the mechanical resonance of the partition material.

Mechanical resonance is a phenomenon that occurs in all physical things, as described briefly in the discussion of infrasound in Chapter 1. To reiterate, the phenomenon of resonance is the amplification of the amplitude of vibration of a specific object that occurs when the object is exposed to high levels of sounds having a specific frequency or set of frequencies related to the physical properties of that object. The resonance region shown below the mass law region in Fig. 4-5 corresponds to the mechanical resonance of the wall. The region above the mass law region corresponds to what is known

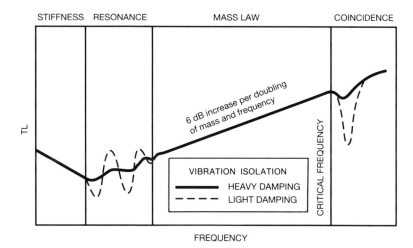

FIGURE 4-5.   Generalized TL spectrum for typical homogeneous partitions. —, Heavy damping; –––, light damping.

as the bending wave resonance of the wall. The resultant drop in TL occurs at what is known as the coincidence frequency. The frequency at which coincidence first affects the TL spectrum is known as the critical frequency. Critical and coincidence frequencies are dependent on the thickness of the material. Table 4-2 shows a range of critical frequencies for common building materials.

Critical frequencies generally decrease at the same rate as the increase in material thickness. In other words, if the material thickness doubles, the critical frequency will be cut in half. Coincidence phenomena can be reduced significantly by using properly designed multilayered partitions. Both mech-

TABLE 4-2   Critical Frequencies of Common Building Materials

Material	Thickness (in.)	Critical Frequency (Hz)
Concrete, poured or block	8	100
Plywood	1/2	1700
Gypsum (drywall)	1/2	3100
Steel, aluminium	1/8	4100
Lead	1/2	4400
Glass	1/8	4900
Plexiglas	1/8	9800

anical and bending wave resonance phenomena can cause significant TL reductions in their respective frequency ranges.

Equation (4-4) can be solved for TL for a homogeneous (having the same material makeup throughout, so that only one τ would apply) partition. If a wall has windows, doors, air gaps, or areas having different materials, Eq. (4-4) can be solved by using a composite τ value defined as follows:

$$\tau_{comp} = (\tau_1 S_1 + \tau_2 S_2 + \cdots + \tau_n S_n)/S \qquad (4\text{-}5)$$

where τ_1, τ_2, and so on, are transmission coefficients for each different partition element and S_1, S_2, and so on, are the surface areas corresponding to the sections of the partition having the same τ subscript number. S is the total surface area of the entire partition.

An example of this calculation would be instructive. Suppose we have a partition wall 10 ft high by 20 ft wide with a 2×4 ft window and a 3×7 ft door built into the partition wall. The individual transmission coefficients are given as:

$$\tau_{wall} = 1 \times 10^{-5} \qquad \tau_{window} = 1 \times 10^{-3} \qquad \tau_{door} = 1 \times 10^{-2}$$

The window area would then be 8 ft², the door area would be 21 ft², and the rest of the wall area would be $200 - 8 - 21 = 171$ ft². Using Eq. (4-5), we would have

$$\tau_{comp} = [(1 \times 10^{-5})(171) + (1 \times 10^{-3})(8) + (1 \times 10^{-2})(21)]/200 = 0.0011$$

Using this τ value in Eq. (4-4), the TL of this partition with the window and door closed would be 30 dB. Using the τ_{wall} value in Eq. (4-4), we see that the TL of the wall by itself, without the window or door, would be 50 dB. Therefore, 20 dB of noise attenuation ability was lost by this wall when the window and door were installed. If the original partition had the window half open, the open portion of the window would have a $\tau = 1$, and solving Eq. (4-5) would yield a $\tau_{comp} = 0.021$, which would correspond to a TL of 17 dB, using Eq. (4-4). Therefore, opening the window half way causes another 13 dB of TL to be lost. As can be seen from this example, the TL of a partition is usually heavily influenced by the part of the partition that has the largest τ value. Small air gaps or doors or windows with low TL values can significantly degrade published TL wall values. Table 4-3 illustrates how small openings in walls can degrade TL performance.

It can be seen from Table 4-3 that the slightest air gap in a partition can significantly degrade the TL of the partition. For this reason, TL values

TABLE 4-3 Transmission Loss Reduction vs. Percent Air Gap[a]

Percent of Wall Area with Air Gap	Resultant TL of Wall (dB)	Resultant Reduction in TL (dB)
0.01	39	6
0.1	30	15
0.5	23	22
1	20	25
5	13	32
10	10	35
20	7	38
50	3	42
75	1	44
100	0	45

[a]Calculations based on original wall TL of 45 dB.

quoted in product specifications literature are valid in practice only if the partition is sealed along its entire perimeter with a soft, nonhardening material such as neoprene or nonhardening caulking. Any hard sealants tend to crack with time and create air gaps for sound leaks. Windows and doors are common sources for sound leaks because of air gaps around their perimeters. Windows must be seated in soft, resilient materials and doors must seal with their frames with soft, resilient materials in order to meet their published specifications. The bottoms of doors should also have dropseals (with soft, resilient materials) that fall onto solid thresholds when closed.

Also important for preserving the TL rating of a partition is the lack of a rigid connection between sides of the partition. Such rigid connections, provided through studs, conduits, ducts, and radiators, provide vibration channels for sound to travel through with minimal attenuation. For this reason, staggering or isolating studs (as is shown in Fig. 4-6) and isolating conduits, ducts, and radiators are necessary to preserve manufacturer TL ratings.

Typical exterior wall designs on buildings provide at least 35 dBA of noise attenuation to interior spaces; however, typical closed windows and doors provide only 15 to 20 dBA of attenuation (the average attenuation of exterior walls with open windows is 10 to 15 dBA). Therefore, unless exterior noise levels are extremely high (above 90 dBA), acceptable interior noise levels depend on the attenuation characteristics of the exterior windows and doors of a building. As mentioned above, multilayered design in windows is typically achieved by separating single panes by air spaces. Noise attenuation increases as the pane thickness and separation distance between panes

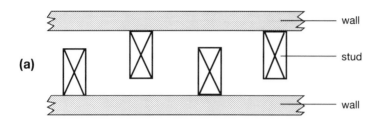

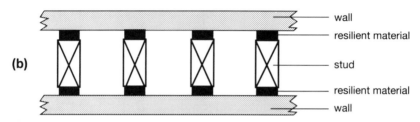

FIGURE 4-6.   Effective methods to increase the TL of a typical studded wall (cross-sectional top view). (a) Staggered studs; (b) studs connected to walls through resilient channels.

increases. When two panes are involved, these are known as double-glazed windows. Laminating windows also provides a multilayered design. Typical double-glazed and laminated windows can provide up to 30 dBA of noise attenuation. Higher attenuation values can be attained by using triple-glazed or specialized double-glazed designs. When these types of windows are required for noise attenuation purposes, they will not be able to be opened to the outside. Therefore, alternate means of ventilation (such as central air conditioning) that would not compromise the noise-attenuating properties of the exterior walls would be required.

Solid-core doors, assumed sealed as stated above, provide 15 to 20 dBA more noise attenuation than do typical hollow-core doors. When more than 30 dBA of noise attenuation is necessary, specialized self-sealing models are available. If possible, a nonsensitive vestibule area between two layers of doors would provide the maximum attenuation possible for practical design.

Sound Transmission Class
A single number rating system for TL covering the human speech frequency range is known as the sound transmission class (STC). Although TL has dB units, the STC has none. The STC is not an average TL value, but a rating

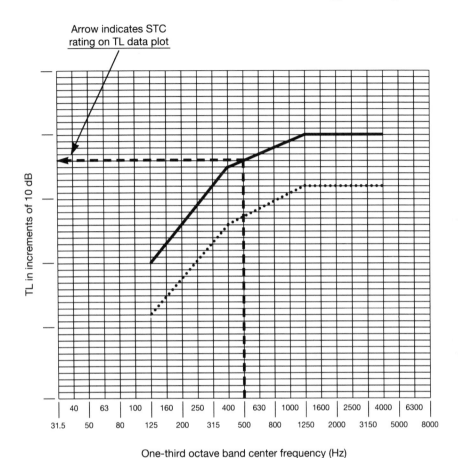

One-third octave band center frequency (Hz)

FIGURE 4-7.   Standard STC curve, according to ASTM E 413-87. The STC values are derived by using the highest level contour (solid line) that satisfies the following criteria: (1) The sum of the differences between individual 1/3-octave TL values and the solid line (below the line) is less than or equal to 32 dB; and (2) the largest difference between an individual 1/3-octave TL value and the solid line (below the line) is 8 dB (shown as the dotted line). The TL value, at 500 Hz, of the highest contour that satisfies these criteria is the STC. (Copyright ASTM. Reprinted with permission.)

based on matching the TL frequency spectrum to a standard curve shape specified in ASTM Standard E 413-87, *Classification for Rating Sound Insulation.* The details of STC derivation are shown in Fig. 4-7. As with NRC values, STC values are useful only when the objectionable frequency range of the source is within the 250- to 2000-Hz frequency range. When human speech is not the only concern, the objectionable frequency range must be identified and, if it is out of the 250- to 2000-Hz range, a partition must be

SOUND PRIVACY	STC	PARTITION DESIGN
Practical Limit	-70-	•Two 6-inch concrete block walls with gypsum on both sides on steel studs with a 4-inch air space between fiberglass and insulation filler.
Complete Privacy (only high level noise heard in low background noise)	-60-	•Two layers of 5/8-inch gypsum on each side of staggered steel studs with a 3-inch cavity and fiberglass insulation filler.
Adequate Privacy (only raised voices heard in low background noise)	-50-	•2 x 4 staggered wood studs with 1/2-inch gypsum on both sides and fiberglass cavity. •6-inch concrete block wall with gypsum on both sides. •Typical building code STC specification between residences.
	-40-	
Some Privacy (voices heard in low background noise)	-30-	•2 1/2-inch steel studs with 5/8-inch gypsum on both sides. •Steel door filled with fiberglass and sealed. •2 x 4 wood studs with 1/2-inch gypsum on both sides. •Sealed solid-core wood door. •Typical double-glazed glass window. •1/8-inch single pane glass window
	-20-	
		•Typical sealed hollow-core wood door. •Typical unsealed hollow-core wood door.
No Privacy (voices heard clearly between rooms)	-10-	•Typical acoustical absorption material.
	-0-	•Open door or window; no partition.

FIGURE 4-8. General STC characteristics for common partitions.

used that is effective in the frequency range of interest. Sound insulation ability in the speech frequency range increases as the STC increases.

Figure 4-8 shows common partitions with their associated STC values. Most building codes require a minimum STC of 45 for partitions between multifamily dwellings. Specific building code requirements are discussed in Chapter 7. Table 4-4 lists TL and STC values for common partitions.

When estimating the STC of a partition having elements with different STC values (e.g., doors or windows), the composite STC can be derived in the same way the composite TL was determined, that is, by using composite transmission coefficients. Composite STC τ values can be derived by solving Eq. (4-6) for τ values for each element, Eq. (4-5) for τ_{comp} values, and Eq. (4-4) (replacing TL with STC):

$$\tau = 10^{-STC/10} \tag{4-6}$$

For example, a 10×8 ft wall having an STC of 45 has a 3×5 ft window with an STC of 25. Solving Eq. (4-6), we obtain $\tau_{wall} = 3.16 \times 10^{-5}$ and $\tau_{window} = 3.16 \times 10^{-3}$. Replacing these values, along with their respective surface areas, into Eq. (4-5), we find

$$\tau_{comp} = \{(3.16 \times 10^{-5}) \times [(10 \times 8) - (3 \times 5)]$$
$$+ (3.16 \times 10^{-3}) \times (3 \times 5)\}/80 = 6.18 \times 10^{-4}$$

Replacing τ_{comp} into Eq. (4-4), with STC instead of TL, we obtain a composite STC of 32.

The derivation of STC values involves using laboratory-condition measurements of TL, according to the methods prescribed by ASTM E 90-90, *Standard Test Method for Laboratory Measurement of Airborne Sound Transmission Loss of Building Partitions.* When measurements are performed on actual installations in nonlaboratory environments, the resultant value is called the field sound transmission class (FSTC). ASTM E 336-90, *Standard Test Method for Measurement of Airborne Sound Insulation in Buildings,* prescribes the measurement methods to be performed. Field transmission loss (FTL) values measured in this manner are then rated against the standard STC curve in ASTM E 413-87 (and Fig. 4-7) to determine the FSTC.

Insertion Loss

Insertion loss (abbreviated IL) refers to the difference in sound levels before and after a noise control device has been inserted into the sound propagation path. IL is most typically used to describe the noise attenuation performance of mufflers, in-duct silencers, and barriers. Insertion loss, rather than TL, is

TABLE 4-4   Acoustic Insulation Performance of Common Partitions[a]

| | Transmission Loss (dB) | | | | | | |
| | Octave Band Center Frequency (Hz) | | | | | | |
Partition Design	125	250	500	1000	2000	4000	STC
Flexible nonlead barrier	15	19	21	28	33	37	27
Quilted blanket	15	19	21	28	33	37	27
Clear vinyl barrier	9	11	16	22	27	32	20
Vinyl acoustic curtain	11	12	18	22	29	34	22
PVC acoustic curtain	12	13	16	21	27	33	21
Leaded vinyl curtain	13	16	20	26	31	35	25
Felt, cotton	4	2	2	2	4	5	3
Glass, 1/8 in.	18	21	26	31	33	22	26
Plexiglas							
1/4 in.	16	17	22	28	33	35	27
1/2 in.	21	23	26	32	32	37	30
1 in.	25	28	32	32	34	46	32
Laminated glass							
9/32 in.	25	28	33	36	35	39	36
1/2 in.	31	34	38	40	37	46	40
3/4 in.	34	36	41	43	41	50	43
Two panes							
1/8-in. glass, $2\frac{1}{4}$-in. air space	13	25	35	44	49	43	37
1/4-in. glass, $2\frac{1}{4}$-in. air space	27	32	36	43	38	51	40
3/16- and 1/4-in. glass, $4\frac{3}{4}$-in. air space	27	39	44	52	54	59	48
Steel roof decking							
22 gauge	29	38	45	48	49	49	45
16 gauge	36	45	52	55	56	56	52
Wood door							
$1\frac{3}{4}$-in. thick, hollow, 1.5 lb/ft^2	14	19	23	18	17	21	19
$1\frac{3}{4}$-in. thick, solid, 4.5 lb/ft^2, gasketed	29	31	31	31	39	43	34
Metal door							
$1\frac{3}{4}$-in. thick, 2.7 lb/ft^2	24	23	29	31	24	40	28
$1\frac{3}{4}$-in. thick, 11.7 lb/ft^2, fully gasketed	26	34	40	48	44	52	43

(*continued*)

TABLE 4-4   **(continued)**

Partition Design	Transmission Loss (dB) Octave Band Center Frequency (Hz)						STC
	125	250	500	1000	2000	4000	
Floor–ceiling							
2 × 8 joists, 16-in. OC, 1/2-in. plywood, 3/8-in. plywood, carpet, pad, 1/2-in. drywall ceiling							37
2 × 10 joists, 16-in. OC, 5/8-in. plywood, 1/4-in. particle board, 1/2-in. parquet wood, 1/2-in. drywall							42
Concrete slab floor, 8 in.	32	38	47	52	57	63	50
Wall							
2 × 4 studs, 16-in. OC, 1/2-in. drywall each side	19	28	35	45	42	44	35
2 × 4 studs, 16-in. OC, 1/2-in. drywall each side, with 2-in. insulation							37
2 × 4 studs, 24-in. OC, 1/2-in. drywall each side, 2-in. insulation							40
2 × 4 staggered studs, 16-in. OC, two layers of 5/8-in. drywall	38	46	49	50	50	59	51
Drywall (1/2 in.) on $3\frac{5}{8}$-in. metal studs, 24-in. OC							39
Weather stripping sealant	22	34	41	41	42	42	40
Concrete masonry wall							
22 lb/ft^2							43
43 lb/ft^2							49
79 lb/ft^2							56
Brick wall							
4 in., 39 lb/ft^2							45
4 in., 1/2-in. plaster, 42 lb/ft^2							50
8 in., 83 lb/ft^2							52
12 in., 117 lb/ft^2							59
Gypsum board							
1/2 in.							28
5/8 in.							29

[a]Note that the values listed are typical of the general types of partitions. For specific applications, it is suggested that one use values associated with the specific materials employed.

Sources: BIA (1988); DuPree (1988); Hedeen (1980); NCMA (1990).

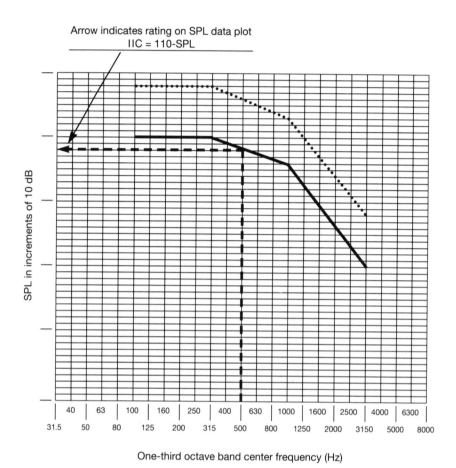

FIGURE 4-9.   Standard IIC curve, according to ASTM E 989-89. The IIC values are derived by using the lowest level contour (solid line) that satisfies the following criteria: (1) The sum of the differences between individual 1/3-octave SPL values and the solid line (above the line) is less than or equal to 32 dB; and (2) the largest difference between an individual 1/3-octave SPL value and the solid line (above the line) is 8 dB (shown as the dotted line). The SPL value, at 500 Hz, of the lowest contour that satisfies these criteria, subtracted from 110, is the IIC.   (Copyright ASTM. Reprinted with permission.)

used for barriers because the effectiveness of a barrier is usually determined by diffraction of sound over it rather than transmission through it. An effective barrier must have a TL at least 10 dB higher than the expected IL value to ensure that the amount of sound transmitted through the barrier is insignificant compared to the amount diffracted over it. Insertion loss calculations are given for partial barriers and mufflers in their sections later in this chapter.

Impact Insulation Class

A single number rating system for the effectiveness of a floor–ceiling combination to reduce sound transmission caused by impact noises, such as walking on floors, is known as the impact insulation class (IIC). Impact insulation class determination uses a curve-matching approach established through ASTM Standard E 989-89, *Standard Classification for Determination of Impact Insulation Class (IIC)*, similar to that used in determining the STC. The details of this method are shown in Fig. 4-9. Because a frequency range lower than that of human speech is normally associated with impact-related structure-borne noise, the IIC curve stresses frequencies below 500 Hz whereas the STC stresses those above 500 Hz. The standard IIC and STC curve shapes match the general frequency trends of the types of noise spectra they are rating. The impact frequency spectrum usually has its maximum levels at low frequencies and sees levels decreasing as frequency increases. The TL spectrum usually has its maximum levels at high frequencies and sees levels increasing as frequency increases.

Impact insulation class values are determined with a standard tapping machine composed of five equally spaced calibrated hammers. The noise level is monitored in the room below while the hammers are tapping on the floor above, as prescribed in ASTM E 492-90, *Standard Test Method for Laboratory Measurement of Impact Sound Transmission Through Floor–Ceiling Assemblies Using the Tapping Machine*. Table 4-5 lists IIC values for common floor–ceiling designs.

TABLE 4-5   Impact Insulation Performance of Common Floor–Ceiling Designs

Floor–Ceiling Design	IIC
Floor–ceiling, 2 × 8 joists, 16-in. OC, 1/2-in. plywood, 3/8-in. plywood, 1/2-in. drywall ceiling	32
Floor–ceiling, 2 × 8 joists, 16-in. OC, 1/2-in. plywood, 3/8-in. plywood, carpet, pad, 1/2-in. drywall ceiling	66
Floor–ceiling, 2 × 10 joists, 16-in. OC, 5/8-in. plywood, 1/4-in. particle board, 1/2-in. parquet wood, 1/2-in. drywall	37
Concrete slab floor (6 in.), 75 lb/ft^2	34
Concrete slab floor (6 in.), 75 lb/ft^2, with 1/2-in. wood fiber board, 44-oz. carpet, 40-oz. hair pad	81

Source: DuPree (1988).

Impact insulation class measurements are based on laboratory conditions and specifications. When such measurements are performed on actual installations in nonlaboratory environments, the resultant value is called the field impact insulation class (FIIC). ASTM E 1007-90, *Standard Test Method for Field Measurement of Tapping Machine Impact Sound Transmission Through Floor–Ceiling Assemblies and Associated Support Structures*, prescribes the measurement methods to be performed. Normalized SPLs measured in this manner are then rated against the standard IIC curve in ASTM E 989-89 (and Fig. 4-9) to determine the FIIC.

Other Rating Methods

Other common rating methods for soundproofing ability include the noise reduction (NR), the noise isolation class (NIC), and outdoor–indoor transmission class (OITC).

The noise reduction (normally denoted NR, but denoted NR_{TL} herein to distinguish it from reverberant field noise reduction discussed in Absorption Calculations, above) is a measure of the soundproofing effectiveness of a single partition between two rooms at a single frequency, defined as the difference between average sound pressure levels (SPLs) in two rooms (with a sound source in one of them) separated by a partition. A definition of TL is given in ASTM E 90-90 as

$$TL = SPL_S - SPL_R + [10 \times \log(S)] - [10 \times \log(A_R)] \qquad (4\text{-}7)$$

where SPL_S is the average SPL in the room with a noise source, SPL_R is the average SPL in the room receiving the sound transmitted through the partition, A_R is the total absorption in the receiving room, and S is the surface area of the partition common to both rooms. Given the definition of NR_{TL} and logarithmic mathematics (from Eq. (2-2)), Eq. (4-7) can be rewritten as

$$NR_{TL} = TL + [10 \times \log(A_R/S)] \qquad (4\text{-}8)$$

The typical range of the $10 \times \log(A_R/S)$ term is -6 to 6. A room, then, with low absorption will lower the TL efficiency (because of noise level increases caused by reverberation) whereas a room with high absorption will raise the TL efficiency of the partition. It should be noted that TL is a property of a partition measured in an ideal laboratory environment whereas the NR_{TL} is a rating of the effectiveness of a partition in an actual installation.

The noise isolation class (NIC) is a single number rating used for sound attenuation effectiveness of a partition derived by using NR_{TL} values, instead of TL values, on the standard STC curve in ASTM E 413-87 (see Fig. 4-7).

Outdoor–indoor transmission class (OITC) is a single number rating system for the sound attenuation effectiveness of a partition, or part of a partition, that separates the outdoor from the indoor environment. This rating is an A-weighted sound reduction value that takes into account the average frequency content of typical outdoor noise sources (such as vehicular, aircraft, and rail traffic) and the characteristic response of specific wall components (i.e., windows, doors, walls, or roofs). Outdoor–indoor transmission class determination is established in ASTM Standard E 1332-90, *Standard Classification for Determination of Outdoor–Indoor Transmission Class.* It can be derived by the measurement methods described in ASTM E 90-90 or ASTM E 966-90, *Standard Guide for Field Measurement of Airborne Sound Insulation of Building Facades and Facade Elements.* ASTM E 90-90 assumes a diffuse field (equal sound energy in all directions) incident on the test panel whereas ASTM E 966-90 allows for provisions in the incident sound field characteristics. The OITC is not as widely used as the STC for rating sound reduction of exterior walls.

Evaluating Vendor Data

Typical vendor sound insulation data specifies TL or STC values for airborne sound reduction. Although the STC value may provide an estimate of noise level reduction in the speech frequency range, the TL spectrum would provide more useful information for noise reduction in the specific frequency ranges of interest. It must be realized that the sound insulation ratings listed by vendors have probably been measured in an ideal testing facility. These values will be realized in the actual installation only if the partition is installed with no air gaps around its perimeter.

Sometimes vendors will state noise insulation only in terms of a percentage decrease in noise levels. This is a deceiving form of advertising that should be exposed for taking advantage of the public ignorance of decibel terminology. Table 4-6 shows the percentage decrease in noise versus the dB decrease in noise level.

TABLE 4-6   Percentage Decrease in Noise vs. dBA Decrease

Percent Noise Decrease	dBA Decrease
25	1
50	3
75	6
90	10
95	13
99	20

TABLE 4-7   Percentage Decrease in Loudness vs.
dBA Decrease

Percent Loudness Decrease	dBA Decrease
50	10
75	20
88	30
94	40
97	50
98	60
99	70

Table 4-6 makes the percentage claims seem much less dramatic. Cutting the noise in half sounds impressive until you realize that this translates to a 3-dBA reduction in noise level, a reduction that is barely noticeable to most people. In terms of the loudness response of our hearing mechanisms, a 10-dBA reduction would correspond to a 50% reduction in loudness. The percentage loudness would then be cut in half with each additional 10-dBA reduction, as is shown in Table 4-7. If the meaning behind cutting the noise in half was that the noise level was reduced from 90 to 45 dBA, as the advertisers would like you to think, they would be making a remarkable claim, comparable to making a lawn mower, heavy truck, or subway train inaudible in most outdoor environments at distances of 10 to 20 ft. Therefore, noise reduction claims in terms of percentages should be avoided unless used in terms of loudness. Transmission loss-related values should be used in rating sound insulation and care should be exercised in installation.

It must be noted that STC and dBA reduction are not identical or interchangeable quantities. Each is derived by a different curve-matching method. For the speech frequency range, the TL dBA value provided by a partition is usually higher than its STC value. Therefore, in general, an STC value provides a minimum, or conservative, estimate of the dBA reduction provided by a partition in the speech frequency range or higher. The opposite can be the case, however, if the noise source of interest has dominant low (below 500 Hz)-frequency components. This is because the STC calculation deemphasizes TL values for frequencies below 500 Hz.

Vibration Isolation Fundamentals
Vibration isolation is a complex field unto itself, about which many volumes have been written, and it is not appropriate to discuss this field in any detail for the purpose of understanding environmental noise pollution. General points to keep in mind are discussed briefly below.

As most of the noise discussed in this book has dominant frequency components above 250 Hz traveling through the air (airborne noise), vibration usually deals with dominant frequency components below 100 Hz traveling through connecting solid materials (structure-borne noise). Structure-borne noise can be generated by low-frequency airborne sources (such as aircraft, trains, or heavy trucks) or by sources that are rigidly attached to buildings (such as mechanical equipment used for ventilation or utilities). Structure-borne noise can travel through an entire building with little attenuation without some basic noise control design considerations. Basically, two principal points must be considered. Resonance conditions must be avoided and the vibration chain must be effectively broken between the source and the rest of the building.

As has been stated several times before, every mechanical system has resonance frequencies associated with it. If any frequencies generated by a source are in the resonance frequency range of building components, vibrations would be carried throughout the building with little attenuation. Therefore, one of the main goals of vibration isolation is to set the vibrational, or forcing, frequency of the connection between the source and the building in a range that would be well out of any resonance frequency ranges involved. Figure 4-10 shows the basic vibration response curve of a system with a single resonance frequency. The transmissibility is the ratio of energy transmitted from the vibrating source to the material to which it is connected. As can be seen from Fig. 4-10, vibration levels are amplified in the resonance frequency region of the curve, and levels are not significantly decreased until the forcing frequency of the source is at least three times that of the resonance frequency.

An analysis tool used to visualize the deformation of an entire structure caused by vibration at resonance frequencies is known as modal analysis. Modal analysis involves dividing system resonances into discrete frequency components and evaluating the deformation patterns caused at each frequency separately. Patterns of displacement similar to standing waves occur at resonance frequencies in solid materials. By visualizing these deformation patterns at each resonance frequency, material properties can be varied on a computer to predict the most effective ways to minimize vibration-related deformations and thus noise and structural damage. Modal analysis is a highly specialized field using complex computer manipulation of vibration data monitored by a narrow-band spectrum analyzer. It should be performed and its data interpreted by experienced personnel only.

It also must be considered that, if the vibrating source is rigidly attached to a structural member of a building, the system to be analyzed would be the combination of the building and the vibrating source. Therefore, the vibrational characteristics of the source when it is attached to something

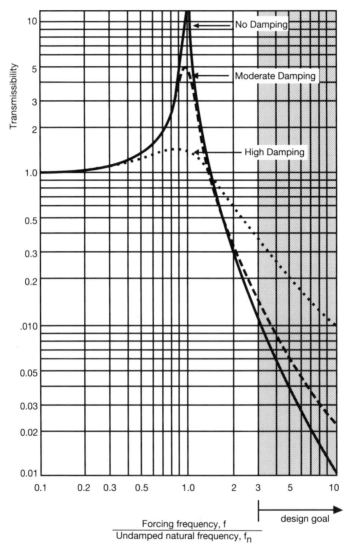

FIGURE 4-10.   Vibration response of the simplest type of system.

would not be the same as those when the source is freely standing. Typical vendor data, if they exist, would deal with the source characteristics only. However, proper treatment of the vibration problem would occur only when the entire system is analyzed. General noise control methods typically used for vibration attenuation are mentioned below in the discussion on noise control in the path.

Noise Control at the Source

Whenever dealing with a noise problem, controlling the noise at the source should be examined first because, if the noise is sufficiently controlled at its source, no other parts of the system need to be examined. This assumes that identifying the noise source is a simple matter. Special monitoring equipment, such as sound intensity systems, can be used to locate noise sources when they are difficult to pinpoint.

Controlling noise at the source would involve maintenance of machinery (such as lubricating, balancing, aligning, securing, and sealing components) or replacing older, noisier equipment with new, quieter models. Proper maintenance is not only important for noise control but minimizes machinery downtime to control operating costs. If feasible, moving or removing the noise source should be considered. Aerodynamic principles should also be considered when dealing with flow-generated noise sources. Avoiding turbulent flows can eliminate significant levels of noise.

Altering the operating conditions of certain machinery may also eliminate a noise problem. The characteristic tones of rotating machinery can be reduced in level and frequency by reducing their rotational speeds. If this is feasible without degrading operation, lowering these tones below 500 Hz adds an extra attenuation that is performed by our hearing mechanisms. If equipment is generating tones in the vicinity of resonance frequencies of the equipment, the extra noise and vibration thus generated can be eliminated by changing operating speeds up or down to move the generated tones out of the resonance range.

The characteristic tones of rotating machinery also generate tones at integer multiples of their characteristic frequency tones, known as harmonics. An early indication of cracks, misalignment, and many other machinery problems that would eventually cause malfunction is a change in sound levels of these harmonics. It may be too late for effective repair or someone may not be present when these anomalous tones become audible. The field of predictive maintenance is based on monitoring these tones and alerting personnel when the tones exceed a preset threshold level. By identifying the frequency or frequencies where limits are exceeded, the component that requires maintenance can be identified and treated before more serious problems arise.

Noise Control in the Path

If it is not feasible or practical to implement noise control measures at the source, the next place to look for treatment is in the path between the source and receiver. This is the most common noise control approach. Treating the

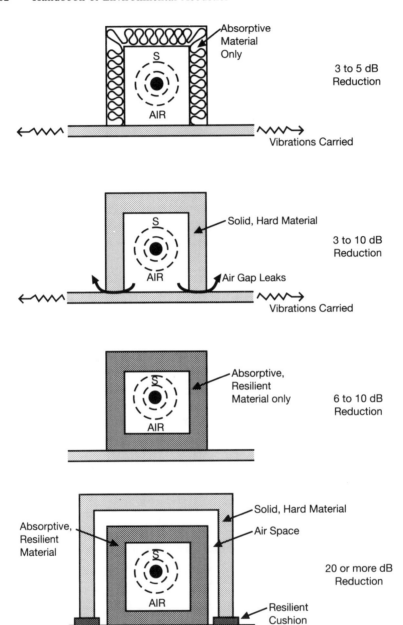

FIGURE 4-11.   Effectiveness of different enclosure designs.

sound path can be divided into two general categories of airborne and structure-borne noise. Controlling airborne noise usually involves some type of barrier (be it an enclosure or partial barrier), muffler, absorptive treatment, or, most recently, active noise cancellation. Structure-borne noise is usually treated by vibration isolation. Most recently, active noise cancellation is also being developed for this purpose.

Enclosure Design

One way to control noise in the path is to completely enclose the noise source of interest. Figure 4-11 illustrates the key points to keep in mind when designing an enclosure. The key points of high TL design must be followed in the design of an effective enclosure. These include completely sealing all openings and breaking all potential vibration channels with soft materials or nonhardening caulking. Absorptive material by itself would not provide an effective enclosure, providing only 3 to 6 dBA of noise attenuation; but a multilayered partition with absorptive material inside the enclosure to minimize reverberation, and thus sound amplification, would be effective in noise attenuation. To be most effective, the enclosure should be sealed airtight; however, if ventilation is required for the source, ducts lined with absorptive material can provide significant attenuation while allowing for proper ventilation. Usually, the attenuation increases as the length of the lined duct and the number of turns in the duct increase. Louvers lined with absorptive material can also be installed at the duct exit to the outside air. These louvers can reduce noise levels by roughly 5 dBA in addition to that attenuated through the ductwork. The duct exit should also, if possible, be facing the direction least likely to cause adverse human reaction. When properly designed and sealed, total enclosure of the source can attenuate noise levels by as much as is necessary (up to a practical limit of 70 dBA).

Figures 4-12 and 4-13 show examples of effective indoor and outdoor enclosures, respectively. Figure 4-14 shows how ventilation can effectively be considered in an enclosure. Other types of enclosures commonly used in industry are shown in Figs. 4-15 and 4-16. Enclosures used for office equipment are shown in Fig. 4-17, and Fig. 4-18 shows the design of acoustic louvers.

A type of enclosure used on piping to control noise is known as lagging material. This is usually wrapped around piping and can reduce the noise level by 20 dBA or more if completely sealed around the pipe in question. Figure 4-19 shows a common lagging installation.

Partial Barrier Design

When complete enclosure of the noise source is not feasible, a partial barrier, one that is open to the air at the top, can be used. It must be noted, however,

FIGURE 4-12.   Indoor enclosure.   (Courtesy of Industrial Acoustics Company, Inc., Bronx, NY. With permission.)

FIGURE 4-13.   Outdoor enclosure.   (Courtesy of Industrial Acoustics Company, Inc., Bronx, NY. With permission.)

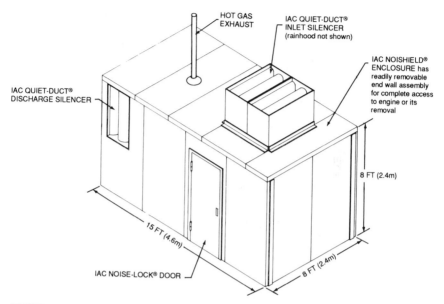

FIGURE 4-14.   Ventilation considerations in an effective enclosure.   (Courtesy of Industrial Acoustics Company, Inc., Bronx, NY. With permission.)

FIGURE 4-15.   Machinery enclosure.   (Courtesy of Kinetics Noise Control, Inc., Dublin, OH. With permission.)

FIGURE 4-16. Machinery enclosure made from heavy curtains for easy access. (Courtesy of Kinetics Noise Control, Inc., Dublin, OH. With permission.)

FIGURE 4-17. Computer printer enclosures. Slots are left open in the back for paper feeding. (Courtesy of Viking Acoustical Corporation, Lakeville, MN. With permission.)

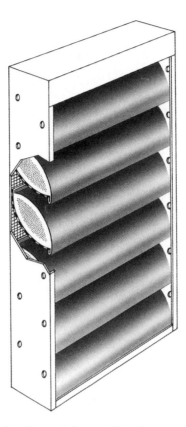

FIGURE 4-18. Cutout view of acoustic louvers. Note the exposure to absorptive material on the inside face.   (Courtesy of Industrial Acoustics Company, Inc., Bronx, NY. With permission.)

that partial barriers have a practical attenuation limit of 10 to 15 dBA caused by diffraction. Partial barriers are typically used outdoors on highways and indoors in open office designs. The noise reduction provided by a partial barrier is usually described in terms of insertion loss (IL) rather than transmission loss. Insertion loss is simply the difference, in decibels, between the noise level without the barrier and the noise level with the barrier. Insertion loss aptly describes the noise reduction provided by inserting a noise control device into the path between the noise source and receiver.

Measurement of the noise reduction performance of barriers is standardized in ANSI Standard S12.8-1987, *American National Standard for Determination of Insertion Loss of Outdoor Noise Barriers,* for outdoor barriers and in ASTM E 1375-90, *Standard Test Method for Measuring the Interzone*

FIGURE 4-19.   Lagging installation (MUFFL-JAC(R) sound attenuating jacketing) at a sewage treatment plant.   (Courtesy of Childers Products Company, Inc., Eastlake, OH. With permission.)

Attenuation of Furniture Panels Used as Acoustical Barriers, for indoor applications.

Many agencies have devised calculation schemes for estimating the effectiveness and best design of partial barriers. All of these are based on acoustic diffraction theory and the chart and illustration shown in Fig. 4-20. The letter N in Fig. 4-20 is known as the Fresnel number, named for the French physicist Augustin Jean Fresnel (1788–1827), who is most famous for his work in diffraction of light in optics. In this case, N describes the sound diffraction over the barrier and w is the sound wavelength.

The derivation of N can become complex if the source and receiver are not at the same elevation. To save time and confusion, the following equation is offered to derive d_1 and d_2 values (from Fig. 4-20):

$$\text{a. } d_1 = \sqrt{(h \times \cos \beta)^2 + (d_3 + h \times \sin \beta)^2}$$

$$\text{b. } d_2 = \sqrt{(h \times \cos \beta)^2 + [d - (d_3 + h \times \sin \beta)]^2}$$

$$(4\text{-}9)$$

where d_1 is the distance between the source and the top of the barrier, d_2 is the distance between the receiver and the top of the barrier, d is the distance

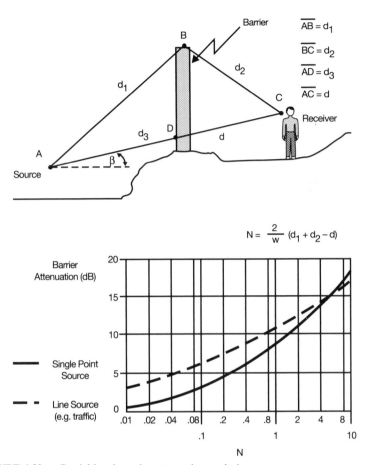

$$N = \frac{2}{w}(d_1 + d_2 - d)$$

FIGURE 4-20.   Partial barrier noise attenuation analysis.

between the source and the receiver, d_3 is the segment of d between the source and the barrier, h is the barrier height above point D in Fig. 4-20, and β is the angle between d and the horizontal; cos is the cosine function and sin is the sine function. When the source and receiver are at the same elevation, $\beta = 0$ and Eqs. (4-9a) and (4-9b) reduce to their appropriate simplified versions in terms of the Pythagorean theorem:

$$\text{a. } d_1 = \sqrt{h^2 + d_3^2}$$

$$\text{b. } d_2 = \sqrt{h^2 + (d - d_3)^2}$$

(4-10)

The Federal Highway Administration (Barry and Reagan, 1978) recommends using different source heights for different vehicles in the barrier calculations. These are 0 ft for automobiles, 2 ft for medium trucks (vehicles with two axles and six wheels), and 8 ft for heavy trucks (vehicles with more than two axles).

Figure 4-20 can be used for barrier insertion loss calculations for typical atmospheric conditions either outdoors or indoors with highly absorptive ceilings. Note from Fig. 4-20 that the effectiveness of a highway barrier would be different for a line of steady traffic (line source) versus a single car or sound source (point source). When dealing with outdoor traffic, an approximation of barrier effectiveness in terms of dBA can be obtained by using 0.9 in the place of the $2/w$ term in the Fresnel number equation in Fig. 4-20. Barrier designers, however, should calculate values for the lowest frequencies of interest. It should also be noted that outdoor barriers may lose their effectiveness beyond distances of 1000 ft because of the effects of atmospheric conditions on sound propagation.

A reflective ceiling above a partial barrier, as can be the case in an office environment, minimizes the effectiveness of the barrier.

The 10- to 15-dBA practical attenuation limit of barriers must be noted, especially when new outdoor noise sources are being introduced to a community. The most common use of outdoor barriers is to reduce highway traffic noise; however, the limits of barrier effectiveness are the same for all sources. It is often thought that a noise source will not be heard if it cannot be seen. This thought, that visual privacy implies acoustic privacy, is usually not the case. Regulations (as discussed in the next chapter) may place limits on noise sources in communities, but, even when these limits are met with noise barriers, noise levels in communities can be much higher than the ambient levels were in the community before the new noise source was introduced. Therefore, people should be fully aware of the effects the new noise will have on a community. When major developments are planned in areas governed by regulations, environmental assessments may reveal the effects of the new noise on the community, as is discussed in the next chapter. However, there are many areas that are not governed by regulations and people are led to believe that a visual barrier will solve any noise problems generated by the new project.

Because the sources typically treated with barriers are outdoors and cannot be enclosed in any practical manner, barriers are usually the only noise control option that would provide any noise attenuation to the receiver. Considering the limits and cost of barriers, it would be more practical and cost effective (considering the hundreds of millions of dollars that have already been spent on highway barriers) to control the noise at the sources instead of providing the partial fix of erecting barriers.

Typical barrier installations are shown in Figs. 4-21 and 4-22. It should also be noted that barriers having absorptive characteristics can provide

FIGURE 4-21.   Highway noise barrier installation.   (Courtesy of Industrial Acoustics Company, Inc., Bronx, NY. With permission.)

FIGURE 4-22.   Railroad noise barrier installation.   (Courtesy of Industrial Acoustics Company, Inc., Bronx, NY. With permission.)

more noise reduction than those without. This owes to the reduction in reflected sound from the barriers. Figure 4-23 shows that absorptive barriers can provide up to 5 dBA more insertion loss than do reflective barriers. This can become even more significant when barriers are parallel to each other on opposite sides of a road, because sound can reflect back and forth between parallel barriers to amplify traffic noise. Some studies have been performed in this area (Caltrans, 1992) that have concluded that, whenever the ratio between the distance between parallel barriers and the height of the barriers

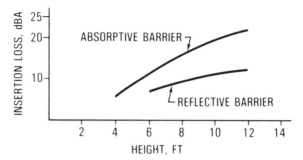

FIGURE 4-23.   Attenuation characteristics of absorptive vs. reflective noise barriers. (Courtesy of Industrial Acoustics Company, Inc., Bronx, NY. With permission.)

is greater than 10 to 1, the amplification caused by the parallel reflective barriers will not be noticeable. Another conclusion of these latest studies is that amplification effects from this scenario have been measured to be less than 3 dBA (Caltrans, 1992; Ridnour and Schaaf, 1987), but more research into this area has been proposed because complaints are being submitted to agencies.

Highway noise barriers are becoming a popular noise control method, with hundreds of miles of noise barriers already constructed and hundreds more in the planning stages. The materials used for noise barriers are typically masonry block, precast concrete, brick, wood, steel, perforated metal sandwiches, plastic, composites, and earth berms (hills). Vegetation (trees, bushes, hedges, etc.) provide practically no significant noise reduction unless they are at least 50 ft tall and 100 ft wide, with dense evergreen foliage extending to the ground (DuPree, 1981). The specific material composition is not as critical as the structural soundness of the barrier because of the practical insertion loss limit associated with the barriers. Any material that is solid (has no air gaps) and will structurally handle the elements can provide at least 15 dBA of noise attenuation. Caltrans (1992) recommends any durable material weighing at least $4 \, lb/ft^2$ for barriers. Highway barriers average 10 to 20 ft in height and cost \$15 to \$20/ft² to erect (FHWA, 1991, 1992).

New barrier materials are constantly being introduced into the marketplace because of the vast market for the product. Innovative highway barrier designs include living walls (made from vegetation, soil, roots, and other natural sources), recycled materials (such as plastic and tires), polystyrene, and transparent walls (made from see-through acrylic).

Average barrier insertion losses are in the 5- to 10-dBA range. To have any effect, a barrier must at least break the line of sight to the receiver. If a sound source can be seen over, through, or to the side of a barrier at a specific location, the barrier will provide virtually no noise reduction to that location. The Federal Highway Administration (FHWA) uses a rule of thumb for estimating highway barrier insertion loss of 5 dBA when the line of sight is broken and 1 dBA more for each additional 2 ft of barrier height above the line of sight. They also suggest that the length of the barrier be four times the distance between the barrier and the receiver (FHWA, 1992). Table 4-8 offers a summary of the practical noise reduction limits of outdoor barriers.

Note that the percentages listed in Table 4-8 are in terms of subjective loudness reduction and not in the form of energy reduction, in accordance with Table 4-7.

Table 4-9 shows results of highway barrier noise measurements for sample installations.

Muffler Design

There are generally two types of mufflers: dissipative and reactive. Dissipative mufflers use absorptive material to reduce noise level and reactive mufflers use the geometric shapes of the device to cause the noise reduction. Figures 4-24 and 4-25 show cutaway sections of typical high-performance dissipative mufflers used in ventilation systems. They have several channels for the sound to be split into where the sound interacts with absorptive material behind perforated metal facings. These types of mufflers provide noise reduction over a fairly wide frequency range.

TABLE 4-8   Outdoor Barrier Noise Reduction Capabilities

Barrier Noise Reduction (dB)	Level of Feasibility	Subjective Loudness Reduction (%)
5	Simple	30
10	Attainable	50
15	Very difficult	65
20	Nearly impossible	75

Source: Simpson (1976).

TABLE 4-9   Measured Insertion Loss of Highway Noise Barriers

Reference Report	Barrier Heights (ft)	Average IL (dBA)
Lindeman (1992)	3 to 22	5 to 15
Armstrong (1980)	16 to 25	9 to 15
Iowa DOT (1983)	9 to 18	10 to 15
Creasey and Agent (1985)	15	6.6 to 7.1 (to 14.4 individual)

FIGURE 4-24.   Cutaway section of ventilation system muffler.   (Courtesy of Industrial Acoustics Company, Inc., Bronx, NY. With permission.)

FIGURE 4-25.   Cutaway section of large, high-performance ventilation system muffler. (Courtesy of Industrial Acoustics Company, Inc., Bronx, NY. With permission.)

Reactive mufflers, on the other hand, are typically effective over limited frequency ranges. For this reason they must be properly matched to the noise source. To avoid leaks from air gaps, mufflers should also be properly fitted and sealed to meet the designed attenuation specifications. Figure 4-26 illustrates the simplest type of muffler, known as an expansion chamber, along with its insertion loss capabilities. When high levels of noise attenuation are required over a broad frequency range, more elaborate mufflers are available, as shown in Fig. 4-27, having many chambers with holes in their walls and absorptive materials. These types of mufflers tend to smooth out the peaks shown for the simple expansion chamber in Fig. 4-26.

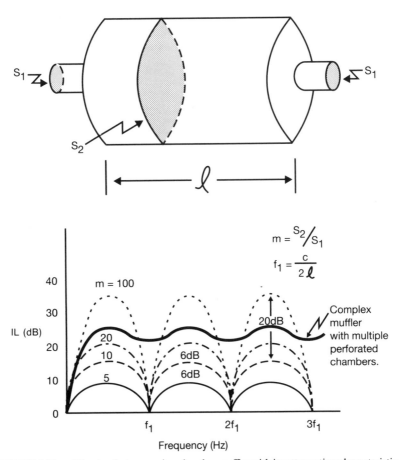

$$m = \frac{S_2}{S_1}$$

$$f_1 = \frac{c}{2\ell}$$

FIGURE 4-26.   The simplest expansion chamber muffler with its attenuation characteristics.

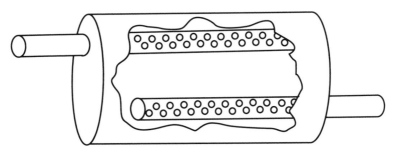

FIGURE 4-27.   Cutaway view of a complex muffler that would level off attenuation characteristics with frequency, as shown in Fig. 4-26.

The principles behind the operation of a simple reactive muffler are the impedance changes caused by the sharp change in cross-sectional area and the acoustic resonance set up inside the expansion chamber. As mentioned in Chapter 1, a sharp change in cross-sectional area in the sound path causes a sharp change in acoustic impedance, which in turn impedes the flow of acoustic energy forward. The sound wave then reflects off one end of the chamber and bounces back and forth within the chamber, without being able to escape. The efficiency of this system decays at standing wave frequencies but can significantly reduce noise levels at other frequencies. These acoustic resonance frequencies are attenuated the least because they counteract the impedance mismatches and their energy thus attains passage through the expansion chamber. A good match is made between muffler and noise source when the acoustic resonance frequencies of the muffler do not match the dominant, objectionable frequencies of the source.

A type of reactive muffler that is effective for absorption of low frequencies is known as a volume or Helmholtz resonator, named after the German physicist Hermann Ludwig Ferdinand von Helmholtz (1821–1894). A Helmholtz resonator is shaped like a jug, open to the air through a narrow neck that feeds into a larger air chamber. The sound energy is trapped inside the chamber because the volume of air in the neck serves as a spring to the volume of air in the larger chamber. The system is therefore like a spring with a mass hanging on it with a resonance frequency governed by the stiffness of the spring and the weight of the mass. For these types of resonators, the highest noise reduction takes place near the resonance frequency of the system unless the walls of the resonator radiate the sound (like a musical jug). Figure 4-28 shows typical absorption characteristics of a Helmholtz resonator.

Helmholtz resonators are used in practice in concrete block wall design. Concrete blocks designed with the Helmholtz resonator idea in mind have narrow slots opening to the air on one side and to an internal chamber on the other side. The frequency range of these blocks is extended by having absorptive material in the chamber inside the block. A cross-section of one of these concrete-masonry-unit Hemholtz resonators is shown in Fig. 4-29. These blocks can effectively reduce noise levels in low-frequency ranges where typical wall treatments do not provide effective reduction. Large reverberant spaces, such as gymnasiums or natatoriums, typically have low-frequency reverberation that can be treated effectively with walls made of these blocks. Figure 4-30 shows an example of such an installation. Figure 4-31 shows an innovative combination of these blocks with a diffusive outer design.

Absorptive Treatment

Acoustically absorptive treatment is usually helpful in long ductwork or, to a limited degree, in reverberant spaces. Inside a reverberant space, having

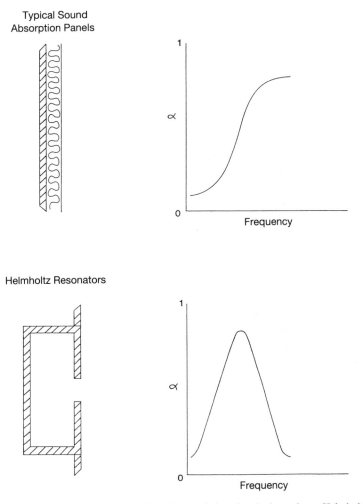

FIGURE 4-28.   Generalized absorption characteristics of typical panels vs. Helmholtz resonators.

hard reflective surfaces for sound to bounce around and amplify levels, strategically placed absorptive materials (with their absorptive properties tailored to absorb the problem source frequencies) can reduce noise levels within that space up to a practical limit of 10 dBA. As stated earlier, bear in mind that the noise levels at the source are not reduced by this method, but only those levels at distances away in the same room.

Examples of different alternatives of absorptive treatment in terms of wall-mounted panels, hanging panels, and hanging resonators are shown in Figs. 4-32 through 4-37. Hanging panels can provide more absorption, using

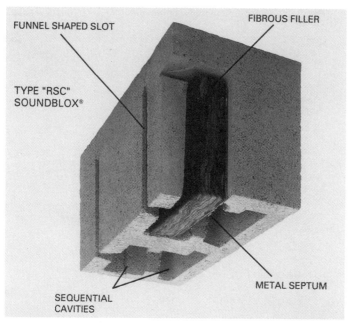

FIGURE 4-29. Cross-sectional view of Helmholtz resonator concrete block. (Courtesy of the Proudfoot Company, Inc., Botsford, CT. With permission.)

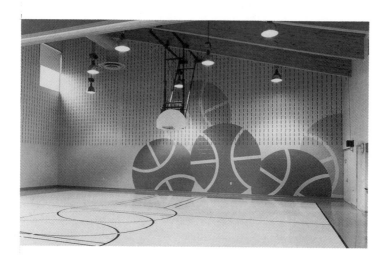

FIGURE 4-30. Helmholtz resonator concrete block installation in a gymnasium. The dark slots in the walls are the openings into the block chambers. (Courtesy of the Proudfoot Company, Inc., Botsford, CT. With permission.)

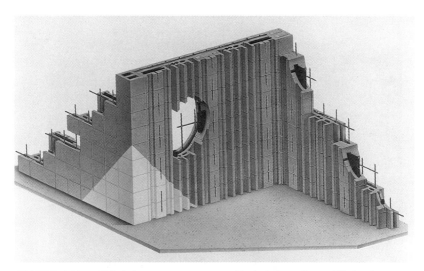

FIGURE 4-31. Helmholtz resonator concrete block design using a diffusive face. (Courtesy of RPG Diffusor Systems, Inc., 651-C Commerce Drive, Upper Marlboro, MD, 20772. With permission.)

FIGURE 4-32. Wall-mounted flat absorptive panels installed in a music room. (Courtesy of Industrial Acoustics Company, Inc., Bronx, NY. With permission.)

FIGURE 4-33.   Wall-mounted absorptive panels with wedges in an office environment to control corner reflections.   (Courtesy of illbruck, inc.; SONEX Acoustical Division.)

FIGURE 4-34.   Hanging flat absorptive panels to control reverberation in a natatorium. (Courtesy of Kinetics Noise Control, Inc., Dublin, OH. With permission.)

FIGURE 4-35.   Hanging absorptive panels with wedges to control reverberation in a gymnasium.   (Courtesy of illbruck, inc.; SONEX Acoustical Division.)

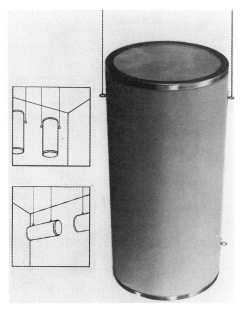

FIGURE 4-36.   Hanging resonator absorber configurations.   (Courtesy of the Proudfoot Company, Inc., Botsford, CT. With permission.)

FIGURE 4-37.   Hanging resonator absorbers installed in a gymnasium.   (Courtesy of the Proudfoot Company, Inc., Botsford, CT. With permission.)

the same amount of material, than attaching panels directly to ceilings because the sound waves can bounce off several absorptive hanging panels instead of one, as when the panels are mounted flush with the wall. Hanging resonators provide a more practical solution to low-frequency problems if a room has been constructed without the built-in wall resonators mentioned above.

Carpet provides the dual acoustic purpose of absorbing airborne and structure-borne sound. Noise reduction coefficient values for standard carpet piles have been measured up to 0.55 without padding and 0.70 with padding (CRI, 1992). In the same study, IIC values were monitored to increase by up to 36 without padding and up to 46 with. Specific noise reduction ratings for carpets are included in Tables 4-1 and 4-5.

When absorption is needed for ceilings (especially for domed ceilings) and it is desired that the material not be seen, acoustically transparent fabric can be used to cover the absorptive material and provide an attractive look. Figure 4-38 shows an example of such an installation.

Most acoustically absorptive materials are not rugged and would disintegrate when exposed to typical weather conditions of precipitation, humidity, wind, and temperature extremes. When absorption is needed for outdoor venues, there are still a few alternatives available for use. An uneven or

FIGURE 4-38.   The use of acoustically transparent fabric to cover absorptive material in a ceiling cove.   (Courtesy of Architectural Fabric Systems, Inc., Westport, CT. With permission.)

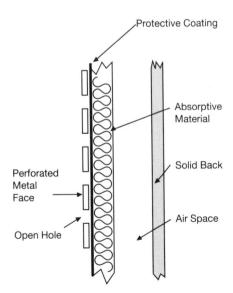

FIGURE 4-39.   Cross-sectional view of a perforated metal sandwich panel.

FIGURE 4-40.   A perforated metal sandwich panel.   (Courtesy of Kinetics Noise Control, Inc., Dublin, OH. With permission.)

porous concrete surface can be effective in absorbing high-frequency sound and diffusing lower frequency sound. Weight and space restrictions usually prohibit this option, leaving the option of using perforated metal sandwiches. A perforated metal sandwich panel has a metal face with holes evenly spaced and sized facing the source of interest. Behind this perforated metal face is a piece of acoustically absorptive material, be it a blanket or board of fiberglass-type material. Behind that material is an air space and then the back face of the panel. This is illustrated in Fig. 4-39. The perforated face protects and supports the absorptive material while providing enough open area to make the absorptive material effective in absorbing the unwanted sound. The ranges of frequencies to be absorbed depend on the size and spacing of the holes in the metal. This must be calculated and tailored to each noise condition.

Perforated metal sandwiches are used in highway barriers as an alternative design for absorptive barriers. They are also commonly used indoors in environments where typical panels would not survive, such as subway stations or industrial environments. Typical porous panels should not be painted because the paint can clog the holes in the panels and render them ineffective for absorption; however, perforated metal sandwiches can be painted to better conform with the aesthetics of an area or room. They should be painted with rollers rather than sprays to avoid clogging the holes. An example of a perforated metal sandwich is shown in Fig. 4-40.

Vibration Isolation

As mentioned above, when a piece of equipment is vibrating and in contact with a solid channel such as a building structural member, this vibration can be carried throughout a building. The most common solution to these types of problems is vibration isolation, in which the vibration source is mounted on some type of resilient material that reduces or eliminates the transmission of the vibrations to the connecting channel. When specific frequencies are a problem, springs that are tuned to increase the forcing frequency of the machinery to at least three times that of the resonance frequency of the system being acted on are the most effective. Using the properly tuned spring is crucial because a matched resonance with a spring isolator can cause a worse vibration problem than originally encountered without the spring.

Figures 4-41 and 4-42 show examples of vibration isolation installations with springs. Springs can also be used to isolate pipe vibrations, as is shown in Fig. 4-43. Another alternative for vibration isolation is the use of resilient pads.

When structure-borne noise is a major concern in a room, a common treatment is known as floating the floor. This means that the floor of a room is structurally isolated from the rest of the building by inserting springs or resilient material between the floor and the structural connections to the rest of the building. Figure 4-44 shows cross-sectional examples of floating floor designs along with their associated noise reduction ratings.

Active Noise Control

All the noise control methods mentioned above are passive methods in that they present an obstacle to the path between the source and receiver. Active noise control changes the source signal directly. Active noise cancellation is a technology that has been emerging, on a practical level, over the past 10 years. Its applications are currently limited but in the future it may prove to be an effective general noise control tool. The theory behind active noise cancellation is that an acoustic pressure wave can be cancelled out by its mirror image. The typical active noise control system is illustrated in Fig. 4-45. A microphone senses the noise and feeds it to a processor that reverses the phase of the signal by 180°. This processor feeds the reversed signal to

FIGURE 4-41.   The use of springs for vibration isolation of heavy machinery.   (Courtesy of Kinetics Noise Control, Inc., Dublin, OH. With permission.)

FIGURE 4-42.   The use of springs for vibration isolation of a vertical pipe.   (Courtesy of Kinetics Noise Control, Inc., Dublin, OH. With permission.)

FIGURE 4-43.   The use of a spring for vibration isolation of a horizontal pipe support. (Courtesy of Kinetics Noise Control, Inc., Dublin, OH. With permission.)

a loudspeaker that emits the new signal to cancel out the original one. An error microphone continually senses corrections needed to update the cancelling signal.

The theory behind active noise cancellation has been around for many years but it has become a reality only in recent years because of the speed capabilities of contemporary electronic circuitry. At present, active noise cancellation is effective in reducing ventilation system noise by 40 dBA or more at troublesome frequencies. The system is usually most effective in a closed or local environment (such as a duct) and on pure tones lower than 500 Hz. Current research in the field of active noise cancellation is targeted at automobile mufflers, aircraft interior silencing, household appliances, automobile interiors, building partitions, and headphones. Figure 4-46 shows the effectiveness of an active silencer and Figs. 4-47 and 4-48 show examples of practical active noise control systems in current use.

In addition to the significant noise reduction provided by active noise

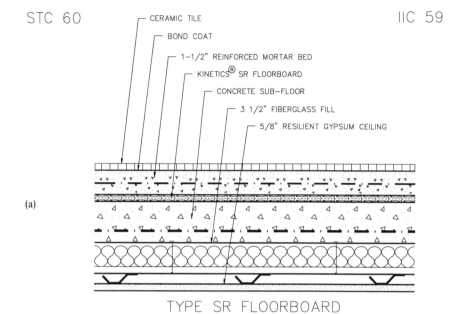

STC 60   IIC 59
— CERAMIC TILE
— BOND COAT
— 1—1/2" REINFORCED MORTAR BED
— KINETICS® SR FLOORBOARD
— CONCRETE SUB—FLOOR
— 3 1/2" FIBERGLASS FILL
— 5/8" RESILIENT GYPSUM CEILING

(a)

TYPE SR FLOORBOARD
WITH 1—1/2" MORTAR BED AND RESILIENT GYPSUM CEILING

STC 50   IIC 59
— VINYL FLOORING
— 1/4" KINETICS® IR UNDERLAYMENT
— 1—1/2" REINFORCED MORTAR BED
— 5/8" KINETICS® SR FLOORBOARD
— 5/8" PLYWOOD SUB—FLOOR
— 3 1/2" FIBERGLASS FILL
— 5/8" RESILIENT GYPSUM CEILING
— 2" x 10" WOOD JOISTS

(b)

TYPE SR FLOORBOARD & IR UNDERLAYMENT
WITH 1—1/2" MORTAR BED AND RESILIENT GYPSUM CEILING

FIGURE 4-44. Examples of floating floor designs, in cross-section, with (a) concrete subfloor and (b) plywood subfloor and wooden joists. Note that the IIC values are the same for each design.   (Courtesy of Kinetics Noise Control, Inc., Dublin, OH. With permission.)

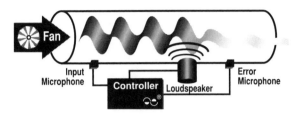

FIGURE 4-45.   Active noise control system for fan noise in a duct.   (Courtesy of Digisonix, Inc., Middleton, WI. With permission.)

cancellation in devices involving flow, such as mufflers, these units introduce none of the pressure differentials that are introduced by typical expansion chamber passive designs. Because of this, engine efficiency can be significantly improved by using the active unit instead of the traditional passive one.

Noise Control at the Receiver

When noise reduction methods are impractical at the source or in the path, treatment at the receiver becomes the last alternative. There are generally three categories of receiver treatment: moving to an alternate location, isolating the receiver, or using hearing protection. Considering that the receivers in the case of environmental noise assessments are usually people who are not to be disturbed by the source in question, control at the receiver is usually an option only when the receiver can be isolated from the source. To be thorough, however, all options are described below.

Changing Source/Receiver Locations

If the location or orientation of the offending noise source or the receiver can be changed in such a way that the offensive nature of the noise source is eliminated or at least minimized, the noise problem may be solved. This can be analyzed by setting the source at a distance away from the receiver at which the source SPL would be the same or less than the background SPL (without the source operating). As a rough estimate, this distance can be approximated by using the following equation:

$$SPL_2 = SPL_1 - [15 \times \log(d_2/d_1)] \tag{4-11}$$

where SPL_2 is the SPL at the new farther location, SPL_1 is the SPL at the original location, d_2 is the distance between the sound source and the new

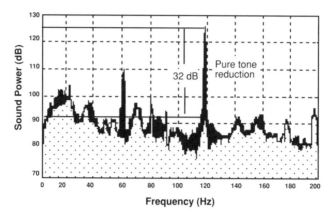

FIGURE 4-46.   Result of active cancellation of fan noise.   (Courtesy of Digisonix, Inc., Middleton, WI. With permission.)

FIGURE 4-47.   An interior view of an active noise-cancelling vehicle muffler.   (Courtesy of Active Noise and Vibration Technologies, Inc.)

farther location, and d_1 is the distance between the sound source and the original location. This translates to a 4.5-dB drop-off per doubling of distance from the source. This takes into account a compromise in sound drop-off rates from the inverse square law to compensate for ground reflection and source characteristics and should be used for approximation only. The actual drop-off rate would depend on the terrain and the source, and can be verified accurately only by measurement. Recall that if the noise source is a line source, such as traffic or a long train, the drop-off rate would be 3 dB per doubling of distance from the source. Once distances away from the noise source exceed 300 ft, the atmospheric absorption factors listed in Chapter 3 must be accounted for.

Changing the orientation of the source or receiver such that the source faces in the opposite direction than the receiver or the receiver is shielded by buildings or natural obstructions can reduce the sound levels reaching the receiver by as much as 10 to 15 dBA. Trees and other foliage tend to be common attempts at noise control, especially for highway traffic sources. As was mentioned above in the discussion on barriers, these natural visual barriers provide negligible noise attenuation, unless they comprise a dense evergreen (as deciduous trees would provide no screening during the winter months without leaves) forest at least 100 ft wide and 50 ft tall. Trees and other foliage should not be considered for noise control purposes.

Isolating the Receiver

Isolating the receiver usually means enclosing the receiver. In terms of mitigating environmental noise impacts, this would mean treating the exteriors of buildings frequented by people to ensure acceptable interior noise levels. This normally involves replacing windows and doors with new ones that provide the attenuation necessary to meet interior noise level goals.

Hearing Protection Devices

Hearing protection devices (abbreviated HPDs) are necessary only when noise levels exceed hazardous limits. Usually, and hopefully, the types of environments requiring HPDs are limited to the workplace, where the Occupational Safety and Health Administration (OSHA) of the U.S. Department of Labor has the authority to ensure that the environments are safe for workers. The OSHA limits and noise regulations are discussed in Chapter 5.

There are, however, nonoccupational activities that involve exposures to noise levels for which HPDs would be appropriate. Examples of recreational activities in which participants should be wearing HPDs are using firearms for hunting or target shooting, and operating power tools such as chain saws or leaf blowers.

FIGURE 4-48.   An active noise control installation in an industrial plant.   (Courtesy of Digisonix, Inc., Middleton, WI. With permission.)

Two types of HPD are generally available to the public: muffs and plugs. Muffs fit over and around the pinna and plugs fit into the opening of the ear canal. Examples of these HPDs are shown in Fig. 4-49.

Effectiveness

The effectiveness of an HPD in reducing noise levels that would otherwise reach the inner ear is usually rated in terms of a single number rating known as the noise reduction rating (denoted NRR). The NRR is defined by the U.S. Environmental Protection Agency (EPA) in Title 40, Part 211 of the Code of Federal Regulations (denoted 40 CFR Part 211). The NRR is expressed in decibels and is a measure of the amount of noise level reduction or attenuation that the HPD would provide to the wearer.

Muffs generally have NRR values between 15 and 27 dB, whereas plugs have NRR values between 10 and 32 dB. When installed properly, plugs can provide the wearer with slightly more attenuation because the ones with high NRR values form to the shape of the ear canal to provide a better seal

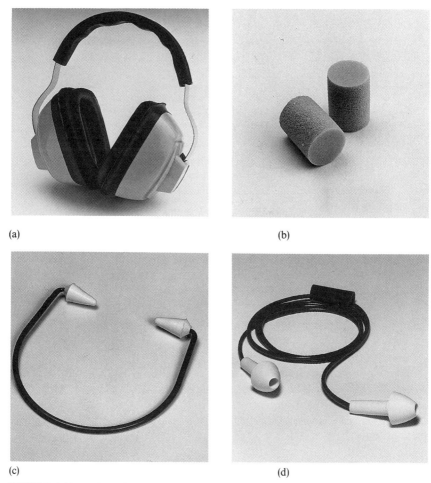

(a)

(b)

(c)

(d)

FIGURE 4-49.   The different kinds of HPDs most commonly used include muffs (a) and several kinds of plugs, including (b) foam formable, (c) semiaural, and (d) premolded. (Courtesy of Cabot Safety Corporation, Southbridge, MA. With permission.)

than muffs can against the skull. The most common types of plugs are foam and flexible. Foam plugs are compressed before being inserted into the ear canal, after which they expand to form a seal. Premolded plugs are also available that are custom fitted to the ear of the wearer by taking a mold of the individual ear canal. A combination of muffs and plugs worn simultaneously can provide an NRR of around 40 dB.

Devices other than plugs or muffs not specifically designed to protect

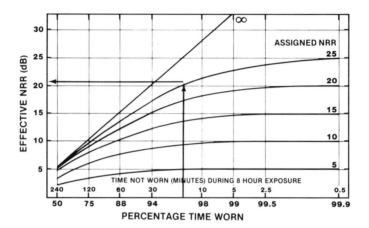

FIGURE 4-50.   The effect of not wearing HPDs for a percentage of exposure times. (From Berger, E.H. 1980a. EARLog #5—hearing protector performance: how they work and what goes wrong in the real world. Sound and Vibration 14(10):14–17. With permission.)

hearing are usually ineffective. Cotton, although used often, provides minimal protection. If HPDs are not available, the next best thing would be to seal the ear canal with a finger.

As is shown in Fig. 4-50, not wearing or removing HPDs, even for short periods of time, can significantly compromise the effectiveness of the HPDs. As is shown in the example, not wearing the HPDs rated at an NRR of 25 dB for just 15 min over an 8-h exposure period would reduce the effective NRR by 4 to 5 dB.

Evaluating Vendor Data

The EPA requires HPD vendors to label HPD packages with the NRR value. Higher NRR values translate to greater noise attenuation to the inner ear, if the HPD is used properly. In terms of actual dBA noise reduction values, it is generally accepted to subtract 7 from the NRR value to approximate the dBA reduction (to account for the inaccuracy of the NRR calculations in terms of frequency distribution). Once this adjusted NRR value is subtracted from the outside noise level, one would know the dBA level to which the inner ear would be exposed. Because of the variability of testing parameters, differences between NRR values of up to 5 dB have little meaning.

There seems to be a large discrepancy between rated and actual HPD effectiveness, especially for plugs. This is mainly due to improper operator insertion. To be inserted properly, the operator must tightly compress the

TABLE 4-10   Noise Control Options

Control Location	Noise Control Option
Source	Maintenance
	Avoid resonance conditions
	Relocate source
	Remove unnecessary sources
	Use quieter models
	Redesign source to be quieter
Path	Enclose source
	Erect barrier
	Install proper muffler
	Install absorptive treatment
	Isolate vibrations
	Use active noise control
Receiver	Relocate receiver
	Enclose receiver
	Use HPDs

plugs between the fingers and insert the HPDs far enough into the ear canal so that they expand to form a tight seal. If they are not inserted far enough or not compressed enough, they will not seal off the ear canal and their effectiveness will be compromised.

A principal concern of people wearing HPDs is their ability to hear speech and warning signals. As long as background signals are at least 80 dBA, HPDs do not seem to impair intelligibility and may actually improve speech intelligibility for background levels above 85 dBA (Berger, 1980b).

As a summary, Table 4-10 is an outline of the most common noise control considerations available to solve a noise problem.

REVIEW QUESTIONS

1. Assume a rectangular room with floor dimensions 30×40 ft and a ceiling height of 15 ft.
   a. If all room surfaces have an absorption coefficient of 0.5 at 500 Hz, what is the total absorption of the room at 500 Hz?
   b. What would be the noise reduction in this room at 500 Hz if the floor were carpeted ($\alpha = 0.4$ at 500 Hz), the ceiling were covered with tile ($\alpha = 0.7$ at 500 Hz), and one of the 30×15 ft walls were covered by a heavy curtain ($\alpha = 0.5$ at 500 Hz)?
2. What would be the change in reverberation time between the conditions of questions 1a and 1b above?

3. A wall 15 ft high and 40 ft wide has a τ of 2×10^{-4} at 500 Hz. What would be the change in TL for the wall if a 1-in. air gap were open along the top of the wall between it and the ceiling?

4. If the room described in question 1b above were on the other side of the wall in question 3 above, what would be the NR_{TL} to the room?

5. A wall 9 ft high and 25 ft wide has an STC rating of 50. An STC 30 door (7 ft high and 4 ft wide) and an STC 27 window (3×3 ft) are installed in the wall. What would be the composite STC of the partition?

6. Derive the STC values for two of the partitions listed in Table 4-4.

7. Name four partitions that have the same critical frequencies.

8. A highway noise barrier is 10 ft tall. Assume all locations are at the same altitude. An idling truck is sitting at a horizontal distance of 50 ft from the barrier. Without the barrier, the measured SPL from the truck at 100 ft is 75 dB at 500 Hz. For the questions below, assume that there is no sound attenuation in addition to that provided by the barrier.

  a. If a person is at ground level 100 ft from the truck, what SPL would he hear with the barrier in place?

  b. If the person were 6 ft tall and standing at 100 ft from the truck, what SPL would he hear with the barrier in place?

  c. If the person were 100 ft from the truck, and standing on an outside deck of a house 20 ft off the ground, what SPL would he hear with the barrier in place?

9. If the idling truck in question 8 above were replaced by a steady stream of moving traffic with a measured SPL of 72 dB at 500 Hz at 100 ft, answer question 8a–c.

10. There are several noisy pieces of equipment (requiring ventilation) in a large reverberant office area. What options are there to control this noise for workers in the area and how effective could each measure be? What options would there be if this equipment were outdoors in a residential neighborhood?

References

Armstrong, R.E. 1980. *Effectiveness of Noise Barriers along the Capital Beltway (I-495) in Northern Virginia.* Washington, D.C.: U.S. Department of Transportation Federal Highway Administration Office of Environmental Policy.

Barry, T.M. and J.A. Reagan. 1978. *FHWA Highway Traffic Noise Prediction Model.* FHWA-RD-77-108. Washington, D.C.: U.S. Department of Transportation Federal Highway Administration Office of Environmental Policy.

Berger, E.H. 1980a. EARLog #5—Hearing protector performance: how they work and what goes wrong in the real world. Sound Vibration 14(10):14–17.

Berger, E.H. 1980b. *The Effects of Hearing Protectors on Auditory Communications.* EARLog #3. Indianapolis, IN: Cabot Safety Corporation.

BIA. 1988. *Sound Insulation—Clay Masonry Walls.* TN-5A. Reston, VA: Brick Institute of America.

Caltrans. 1992. *California Noise Barriers.* State of California Business and Transportation Agency, Department of Transportation.

CISCA. 1984. *Acoustical Ceilings: Use and Practice.* Elmhurst, IL: Ceilings & Interior Systems Construction Association.

Creasey, T. and K.R. Agent. 1985. *Effectiveness of Traffic Noise Barrier on I 471 in Campbell County, Kentucky.* UKTRP-85-15. Lexington, KY: Kentucky Transportation Research Program.

CRI. 1992. *The Carpet Specifier's Handbook,* 5th ed. Dalton, GA: The Carpet and Rug Institute, pp. 40–44.

DuPree, R.B. 1981. *Evaluation of Noise Reduction of Building Facades and Outdoor Noise Barriers.* Berkeley, CA: Office of Noise Control, State of California Department of Health Services.

DuPree, R.B. 1988. *Catalog of STC and IIC Ratings for Wall and Floor/Ceiling Assemblies.* Berkeley, CA: Office of Noise Control, State of California Department of Health Services.

FHWA. 1991. *Highway Traffic Noise Barrier Construction Trends.* Washington, D.C: U.S. Department of Transportation Federal Highway Administration Office of Environment and Planning, Noise and Air Quality Branch.

FHWA. 1992. *Highway Traffic Noise Analysis: Reasonableness and Feasibility of Abatement.* Washington, D.C: U.S. Department of Transportation Federal Highway Administration Office of Environment and Planning, Noise and Air Quality Branch.

Hedeen, R.A. 1980. *Compendium of Materials for Noise Control.* Cincinnati, OH: U.S. Department of Health, Education, and Welfare, Public Health Service, Centers for Disease Control, National Institute for Occupational Safety and Health.

Iowa DOT. 1983. *An Iowa Noise Barrier: Sound Levels, Air Quality and Public Acceptance.* Iowa Department of Transportation Planning and Research Division.

Lindeman, W. 1992. *Noise Barrier Status Report.* Tallahassee, FL: Florida Department of Transportation.

NCMA. 1990. *Sound Transmission Class Ratings for Concrete Masonry Walls.* TEK 69B. Herndon, VA: National Concrete Masonry Association.

Ridnour, R. and D.V. Schaaf. 1987. *Effectiveness of Parallel Noise Barriers—An Iowa Study.* FHWA Work Order No. DTFH71-83-3610-IA-12. Ames, IA: Iowa Department of Transportation Planning and Research Division.

Simpson, M.A. 1976. *Noise Barrier Design Handbook.* FHWA-RD-76-58. Washington, D.C.: U.S. Department of Transportation Federal Highway Administration Office of Research.

For Additional Information on Specific Topics

Beranek, L.L. 1988. *Noise and Vibration Control,* revised ed. Washington, D.C.: Institute of Noise Control Engineering.

Beranek, L.L. and I.L. Vér. 1992. *Noise and Vibration Control Engineering.* New York: John Wiley & Sons.

Egan, D.M. 1988. *Architectural Acoustics.* New York: McGraw-Hill.

Harris, C.M. 1991. *Handbook of Acoustical Measurements and Noise Control*, 3rd ed. New York: McGraw-Hill.

Schaffer, M.E. 1991. *A Practical Guide to Noise and Vibration Control for HVAC Systems*. Atlanta, GA: American Society of Heating, Refrigerating, and Air-Conditioning Engineers, Inc.

Schultz, T.J. 1986. *Acoustical Uses for Perforated Metals: Principles and Applications*. Milwaukee, WI: Industrial Perforators Association.

5

Noise Regulations, Guidelines, and Ordinances

Noise regulations and guidelines of all major agencies that deal with noise issues are discussed in this chapter for information and comparison. Their criteria can be used as the best guidelines and references for developing any new noise criteria. Before discussing the details of specific guidelines and regulations, it would be most appropriate to describe the types of noise sources that are typically regulated.

TYPES OF NOISE TYPICALLY REGULATED

The principal types of noise sources affecting most community environments can be divided into three categories: mobile source, stationary source, and intermittent/temporary. Each is discussed below.

Mobile Source Noise

Mobile sources, according to the dictionary, are sources that move with respect to some point of reference. In terms of the community environment, mobile noise sources would be sources that move in reference to a noise-sensitive location. These, most typically, would be transportation-related vehicles such as automobiles, buses, and trucks (the vehicular traffic sub-category), trains, and aircraft. Each of these types of sources has its own distinctive noise signature and, consequently, set of noise assessment descriptors associated with it. The details of these signatures and descriptors are discussed in the following sections.

Vehicular Traffic

Vehicular traffic includes automobiles, buses, and trucks. The types of noise generated by each of these vehicles have similar characteristics in that each type can be divided into two independent subsources of engine and tire noise. The levels of each of these subsources is more dependent on the engine than on vehicle speed; however, trucks and cars are different in their level and spectral characteristics. Buses and trucks are similar in their respective noise characteristics.

In general, automobiles produce noise levels that are independent of vehicle speed but vary with the logarithm of the engine speed. With changing gears, the noise levels tend to increase in a sawtooth kind of pattern as vehicular speed increases. The interaction of the road surface with tires also generates noise that increases with vehicle speed. At vehicular speeds below 30 miles per hour (mph), the typical automobile noise spectrum is dominated by engine noise. At speeds higher than 30 mph, the automobile noise signature is composed of a combination of lower frequency engine noise and higher frequency tire noise. The engine and tire noise above vehicular speeds of 30 mph are comparable in noise level.

Although the noise generated by buses and heavy trucks is also composed of engine and tire noise, the characteristic noise signature is different from those of automobiles. This is because tire noise tends to dominate the noise signature at vehicular speeds above 30 mph in trucks and buses. Payload normally does not significantly affect noise levels because increased payload usually results in decreased vehicular speed and the effects cancel each other out.

The Federal Highway Administration (FHWA) has established relationships between vehicle speed and noise level, in dBA, at 50 ft (Barry and Reagan, 1978) as follows:

$$\text{a. } 38.1 \times \log(s) + 5.5 \quad \text{for automobiles}$$

$$\text{b. } 33.9 \times \log(s) + 23.4 \quad \text{for medium trucks} \qquad (5\text{-}1)$$

$$\text{c. } 24.6 \times \log(s) + 43.6 \quad \text{for heavy trucks}$$

where s is the vehicle speed in miles per hour (with a minimum of 30 mph), automobiles include all vehicles with two axles and four wheels, medium trucks include all vehicles with two axles and six wheels, and heavy trucks include all vehicles with three or more axles. These relationships are based on the mean of noise levels monitored without propagation or traffic flow effects. At different speeds, they translate to the values listed in Table 5-1.

A common designation of vehicular traffic volume for environmental noise assessments is the passenger car equivalent (PCE). The PCE value is the

TABLE 5-1   Federal Highway Administration Noise Level[a] vs. Vehicle Speed

Speed (mph)	Vehicle Type					
	Automobile		Medium Truck		Heavy Truck	
	SPL	PCE[b]	SPL	PCE	SPL	PCE
30	62	1	73	15	80	65
45	68	1	79	12	84	38
55	72	1	82	11	86	29

[a] In dBA at 50 ft.
[b] Passenger Car Equivalents

Source: Adapted from Barry and Reagan (1978).

traffic volume, in terms of automobiles only, that would generate the same noise level as the actual traffic volume, including heavier vehicles. The PCE designation makes analysis methods simpler and more consistent than would be the case if the actual vehicle mixes were included. According to the FHWA relationships stated above, medium trucks correspond to 11 to 15 PCEs with vehicle speed (See Table 5-1). Heavy trucks are more dependent on speed, corresponding to 65 PCEs at 30 mph, 38 PCEs at 45 mph, and 29 PCEs at 55 mph. In other words, 1 heavy truck travelling at 30 mph would, on the average, generate the same noise level at 50 ft as 65 automobiles passing the same location at the same time and speed.

Rail Operations

The principal noise sources of rail systems are the interaction between wheels and rails, the propulsion system of the railcars, auxiliary equipment (ventilation and horns), and, in the case of high-speed trains, aerodynamic noise. The dominant cause of railcar noise over most of the typical speed range is interaction between the wheels and rails. When traveling on straight (also called tangent) tracks, wheel–rail noise is generated by the roughness of the wheel and rail surfaces. Railcars traveling on smooth wheels and smooth continuous welded (jointless) rail emit a steady broadband (i.e., no dominant pure tone frequency components) rolling noise. If rail joints exist and are not aligned or if rails and/or wheels are not smooth, noise levels may be up to 15 dBA higher (at 100 ft) than they would be with proper maintenance.

When railcars travel on curves having radii less than about 350 ft, the dominant noise emitted is a high-pitched squeal or screech. This is usually caused by metal wheels sliding on the rail and scraping metal on metal when a train negotiates a curve. The sliding occurs because of the design of the railcar as the wheels are fixed in location with parallel axles. Therefore the wheels cannot turn around curves and, because the wheels on the outside rail of a curve would have to travel farther than the wheels on the inside rail while on a fixed-radius axle, the outside wheels would have to slide on the track around a curve. New trains are being developed with axles that flex to adapt to such curves and thus reduce this noise; however, most conventional trains do not have such designs.

Electrically powered railcars have additional noise sources related to the propulsion system, including the traction motor, reduction gears, traction motor air-cooling system, and the ventilation system. Another noise source that may be of concern is the horn, which typically generates maximum noise levels of 85 to 90 dBA at 100 ft.

Aerodynamic noise is generated as a result of rapidly fluctuating pressures in the turbulent air on or near the surface of a moving train. The noise levels associated with aerodynamic generation are logarithmically proportional to train speed, and become significant only at speeds above roughly 150 mph for smoothly shaped trains. This noise is independent of whether or not the train is contacting a track, so, at high speeds, all trains would produce similar noise levels dependent only on the aerodynamic smoothness of the train.

When a railcar is traveling on an elevated structure, such as a bridge or elevated guideway, noise levels can be as much as 20 dBA higher than those generated by railcars traveling on tracks at grade. This is primarily caused by radiation of sound from vibrating components of the elevated structure.

Aircraft Operations

The principal noise sources of conventional aircraft (airplanes and helicopters) are the propulsion system and aerodynamic noise. There are generally three types of engines in use on contemporary airplanes: turbojet, turbofan, and propeller. In the turbojet and turbofan models, the dominant noise source is the exhaust, generating the characteristic low-frequency roar of the jet engine. Propeller aircraft have combinations of engine exhaust noise and propeller noise, with the propeller component usually dominating. Propeller noise is typically composed of integer multiples of discrete frequency peaks proportional to the propeller blade rotational speed and the number of propeller blades, just as for fans. This produces the typical whining sound of propeller-driven aircraft.

Aerodynamic noise, as for high-speed rail noise, is generated by airflow around the fuselage, cavities, control surfaces, and landing gear of the airplanes. Aerodynamic noise is usually dominant only at frequencies above 600 Hz during cruise conditions. Conditions during takeoff and landing normally cause the propulsion system noise to be dominant over the aerodynamic component.

The dominant sources of helicopter noise are the engine and rotor systems.

The engine noise is similar to that discussed for airplanes, but on a smaller scale. Rotor noise is periodic in nature and is characterized by low-frequency slaps or cracks caused by the sharp variations in pressure encountered by the rotating rotor blades as they pass through the aerodynamic wake produced by each adjacent blade. As for propeller noise, the frequency of the rotor noise is proportional to the rotational speed and the number of blades in the rotor system. The frequencies associated with typical helicopter rotor noise are lower than those for propellers because the rotor blades are usually much larger and blade speeds are lower for helicopters than for airplanes.

In typical flyover conditions, airplanes exhibit higher frequency domination during the approach and lower frequency domination after passing over. This is the case because the fan and compressor sections are in front of the jet engines, radiating their characteristic higher frequency components while the jet exhaust is at the back of the engine, emitting its characteristic lower frequency noise. The characteristic blade slap generated by helicopter rotor systems usually dominates the noise signature in the approach of a flyover and the engine noise usually dominates the signature after the helicopter has flown overhead and is flying away. In all cases, aircraft flyover noise maximizes in level when the aircraft is approximately overhead, when all noise components are the closest distance to the receptor. Variances in this behavior are caused primarily by atmospheric conditions.

Stationary Source Noise

Stationary noise sources, as the name implies, do not move with respect to a noise-sensitive location. Typical stationary noise sources of concern would be machinery or mechanical equipment associated with industrial and manufacturing operations or building ventilating systems. Also included in this category would be crowds of people within a defined location, such as children in playgrounds or spectators attending concerts or sporting events. The basic characteristics of these sources are described below.

Mechanical equipment generally includes machinery used for industrial purposes such as motors, compressors, boilers, pumps, transformers, condensers, generators, cooling towers, and ventilating equipment. Among the many different types of machinery used in industry, there are common noise-generating methods associated with most of them. These are mechanical (through gears, bearings, belts, fans, or other rotating components), aerodynamic (through air or fluid flow), and magnetic (through magnetostriction or periodic forces between rotors and stators).

Mechanical noise sources are usually generated by gears, bearings, belts, rotors, or fans rotating and moving within a machine. Much machinery noise is caused by factors that could be eliminated with proper maintenance. These factors include imbalance, instability, improper lubrication, uneven surfaces, cracking, excessive dirt, loose parts, and leaks. Maintenance problems can elevate noise levels up to 20 dBA higher than those that would be emitted from properly maintained machinery. The higher levels of noise associated with maintenance problems can be used as a valuable tool to diagnose machinery faults before the equipment self-destructs. As mentioned in Chapter 4, a branch of vibration analysis known as predictive maintenance has been established over the past 10 years to deal effectively with such potential disasters by monitoring equipment for vibration levels and alerting personnel when levels exceed acceptable thresholds.

Assuming proper maintenance, mechanical machinery noise is usually characterized by discrete mid- to high-frequency tones proportional to the rotational speed of the machinery components. These tones are usually caused by friction, vibration of components, and aerodynamic flow generation. Even when large machinery is properly maintained, noise levels can exceed 100 dBA within 10 ft of the equipment.

Aerodynamic noise sources involve fluid (usually, but not always, air) flow through, around, and out of machinery. Noise usually becomes an issue when the fluid flows through a restrictive, unsmooth path and turbulence is generated. All rotating equipment generates fluid flow by moving fluids as rotating parts operate. When blades are rotating at a constant rate, especially in an enclosure such as a duct, pure tones are usually generated above a certain frequency (known as the cutoff frequency) that are integer multiples (in frequency) of the product of the blade rotational speed and number of blades in the rotor system. Turbine motors and engines have collections of stationary blades set near the collections of rotating blades to smooth out flow for greater efficiency. The interaction of flow coming off of the rotating blades and traveling through the stationary blades (an occurrence known as rotor–stator interaction) also generates tones characteristic of the system. In machinery having several rotating components, it is common that each component radiates its own distinctive set of tones. Changes in the characteristics of these tones is also a valuable tool for targeting problems in machinery diagnostics.

Boilers and steam turbines have liquids and steam flowing through them at high speeds, generating a hissing noise or roaring noise that can exceed 100 dBA within 10 ft. These sources tend to be less pure tone in nature than the rotating machinery noise. Hissing is usually caused by leaks in the system, where high pressures of fluid or gas are exposed to small openings, creating a large amount of flow turbulence that results in noise.

Ventilating systems usually have fans that generate tones at high operating speeds. These tones, in addition to broadband noise generated by the turbulence in flow through unsmooth ducting elbows and discontinuities, can propagate through a building and produce noise in rooms far away from the original source. Systems ignoring smooth flow designs can generate up to 30 dBA higher noise levels than systems that use smooth flow design. Outdoor street level and rooftop air conditioning units usually have fans and condensers that can generate airborne noise to adjacent buildings. If not isolated from the building structure by properly tuned springs or resilient materials (as discussed in Chapter 4), ventilating systems and other machinery can also generate structure-borne noise that may be sensed throughout a building and possibly a neighborhood.

Magnetic noise sources usually exist in electric motors, generators, lighting fixtures, and transformers. In electric motors, magnetic forces are caused by mechanical and electromagnetic properties of the rotor–stator assembly. In generators, strong magnetic forces may deform stators, causing vibration and consequent noise radiation. In fluorescent or mercury lamps and transformers, the phenomenon of magnetostriction (the change in dimension of a material when magnetized) causes their characteristic humming noise. Because transformers usually use currents in the 60-Hz frequency range, the magnetized iron in the transformer vibrates at the same rate and at integer multiples (harmonics) of it. Current and magnetic characteristics for lighting fixtures and transformers are such that higher harmonics of the line frequency are excited and their highest noise levels are in the speech frequency (500- to 2000-Hz) range. This explains the characteristic whine associated with fluorescent lighting and transformers.

People are not usually thought of as stationary sources; however, when they are contained in a defined area under special conditions, such as children in playgrounds or spectators at outdoor sporting events or concerts, they can be treated in that respect and can cause annoyance in communities. The human voice has roughly 500 to 2000 Hz as its dominant frequency range. Children's voices are slightly higher in pitch than those of adults; however, they still fall into this frequency range. When many voices are combined the frequency content smooths out to become broadband in nature over the 500- to 2000-Hz range. Instantaneous crowd noise levels at outdoor events can exceed 100 dBA. Measured crowd noise levels are discussed in Chapter 6.

Intermittent/Temporary Sources

The most common intermittent/temporary noise source is construction activity. Construction noise sources are composed of both mobile sources (e.g., trucks and dozers) and stationary sources (e.g., compressors, pile drivers, and power tools). These are listed under a separate category because they are temporary and limited in their duration of use. This duration limit must be taken into account when determining the noise assessment of construction activities. Typical mobile and stationary sources are considered separately because they are assumed to be a permanent component of the environment.

There is a wide variety in size and type of construction equipment. They can generally be divided into the categories of continuous and impulsive/impact sources.

Continuous sources include motor-driven equipment used for excavation (e.g., dozers, earth drills, loaders, and backhoes), materials-handling equipment (e.g., cement mixers and cranes), and stationary equipment (e.g., compressors, generators, and pumps). The exhaust noise associated with these engines is usually the dominant noise source of each piece of equipment; however, noise from intakes, cooling fans, and transmissions can also contribute to the emitted noise levels.

Impulsive/impact equipment includes devices such as pile drivers, pavement breakers, jackhammers, and small hand-held power tools (e.g., riveters and impact wrenches). In conventional pile drivers, the principal noise source is the impact of the hammer striking the pile and engine sources are usually secondary. When noise is a concern and soil conditions are acceptable, augered or vibratory pile drivers can be used. The dominant noise sources for pneumatic tools are the exhaust and the impact of the tool bit against the materials they hit. Recent noise measurements of common construction sources are listed in Chapter 6.

Other intermittent/temporary noise sources include animal sounds and everyday activities that are normally handled on a local level and are discussed in Chapter 6.

ENVIRONMENTAL NOISE ASSESSMENTS

There are generally two categories of environmental noise assessment. One predicts the effects of a proposed project or noise source on noise-sensitive locations and the other deals with the effects of existing noise sources on noise-sensitive locations. When large-scale, noise-generating projects are proposed, the rigorous process of environmental assessment and environmental impact statements is usually required. Effects on existing locations are usually handled on a local level with ordinances, depending on the noise sources. The analyses of these two categories of assessment can overlap, and therefore the rest of this chapter is organized in terms of governing agencies rather than analysis methodologies.

In an environmental noise assessment of a new project to be built, a noise issue is identified when a noise-sensitive location would be affected by project-generated noise or when a project-related, noise-sensitive location would be affected by the existing noise in the area around the project site.

These points, along with defining a noise-sensitive location, are discussed below.

Noise-Sensitive Locations

A noise-sensitive location is usually considered to be a defined area where human activity may be adversely affected when noise levels exceed predefined thresholds of acceptability or when noise levels increase by an amount exceeding a predefined threshold of change. These locations can be indoors or outdoors. Indoor noise-sensitive locations would include, but would not be limited to, locations such as residences, hotels, motels, health care facilities, nursing homes, schools, houses of worship, court houses, public meeting facilities, museums, libraries, and theaters. Outdoor noise-sensitive locations would include, but would not be limited to, residential yards, parks, cemeteries, outdoor theaters, golf courses, zoos, campgrounds, and beaches. Land use and zoning maps are usually helpful in initially targeting noise-sensitive locations; however, inspection of the area in question firsthand is the only way to identify all noise-sensitive locations that may be affected by a proposed project or existing noise source.

Project-Generated Noise at Existing Noise-Sensitive Locations

Existing noise-sensitive locations potentially affected by project-generated noise are usually identified according to the specific project-generated noise source. If a proposed project would add any of the following:

1. vehicular (especially truck or bus) traffic to existing or new thoroughfares that pass by such locations
2. rail activity to existing or new rail lines within 1000 ft of such unshielded locations
3. aircraft flying through existing or new flight paths over such locations, causing them to potentially be within an L_{dn} 65 contour
4. a stationary source within 1000 ft of such unshielded locations
5. construction equipment and operations within 1000 ft of such unshielded locations

with respect to the noise-sensitive locations listed in the section above in a typical urban or suburban area, these locations would typically be identified as noise-sensitive locations potentially affected by the project. The distances listed above may increase for rural or quiet suburban areas, depending on atmospheric and topographical conditions. Such cases should be evaluated in terms of their specific conditions.

Creation of a New Noise-Sensitive Location

If a proposed project would create any of the types of locations potentially identified as being noise sensitive in any of the five situations described above or if the new location would be

1. adjacent to an active highway
2. within 1000 unshielded feet of an active rail line
3. within an aircraft-generated L_{dn} 65 contour
4. within 1000 unshielded feet of a stationary source
5. within 1000 unshielded feet of construction equipment and operations

these locations would be identified as noise-sensitive locations potentially affected by the project.

Federal and state regulations usually handle these types of noise sources. Any noise sources not identified above are usually handled on a local level.

Environmental Impact Statements

The National Environmental Policy Act of 1969 (Public Law 91-190, as amended by Public Laws 94-52 and 94-83; denoted NEPA) established that federal agencies must analyze the environmental impacts caused by proposed projects in the form of environmental impact statements (EISs) and established the Council on Environmental Quality (CEQ). The CEQ, in 1978, published regulations for implementing the so-called NEPA process in Title 40 of the Code of Federal Regulations, Parts 1500 to 1508 (40 CFR Parts 1500–1508).

If it is clear that no significant impact would result from a proposed project, a categorical exclusion (according to 40 CFR Part 1508.4) may be issued, in which case no other documentation would be required for the project in question. An environmental assessment (EA), as defined in 40 CFR Part 1508.9, is a report that is submitted to determine whether significant (to be defined for each situation) impacts would be expected from a proposed project. If it is determined through the screening analysis of an EA that no significant impacts would result from the proposed project, a finding of no significant impact (denoted FONSI), according to 40 CFR Part 1508.13, is issued. Otherwise, a notice of intent (according to 40 CFR Part 1508.22) is issued and an EIS must be drafted with detailed analyses.

An EIS is a formal document in a prescribed format that describes all potential adverse effects (impacts) that can be expected from the construction and operation of a proposed new project. The potential impacts are derived by using criteria established on local, state, and federal levels. The disciplines typically examined in an EIS include (where applicable) land use and zoning, socioeconomic conditions, open space and recreational facilities, cultural resources, visual quality, neighborhood character, natural resources, hazardous materials, waste management, energy consumption, transportation, air quality, construction, noise, project alternatives, and mitigation.

As mentioned above, an EIS is required to be prepared only after initial screening analyses prove that it would be necessary to perform a detailed analysis because of the likelihood of project-generated impacts. The extent of these impacts could be determined only by this detailed level of analysis in each discipline.

The typical EIS is composed of an executive summary of the document, a description of the proposed project, and technical analyses in each of the applicable disciplines mentioned above. The conditions usually analyzed are the existing conditions (at the time of the study), the conditions in the planned first year of operation of the proposed project if the project were not built (known as the No Build condition), and the conditions in the planned first year of operation of the proposed project if the project were built (known as the Build condition).

A clear indication of the effect a proposed project would have on the given environment can be provided by a comparison between No Build and Build conditions, although there are agencies that compare existing to Build conditions for their impact criteria. The problem with the latter type of analysis is that it does not represent, unless there are minimal environmental changes other than those caused by the proposed project, the effect to the environment of the proposed project alone. Other projects built between the existing and Build years can change the environmental factors enough to cause an impact (when added to the factors contributed by the proposed project) that would not have been caused by the proposed project alone. In this case, the proposed single project that causes cumulative impacts to become significant would be blamed for the entire impact and penalized accordingly. Comparing No Build to Build conditions would reveal the extent of the impact contribution for each individual project. If a significant impact is not caused by a single proposed project but is caused cumulatively when combined with others, it can then be decided how to handle mitigation by apportioning the responsibility to all projects that contribute to the significant impact.

The EIS drafting process normally occurs in several stages. A draft EIS (DEIS) is usually published first and made available to the public and agencies for their comments for a specified time period. Following that time period, the comments must be addressed by the DEIS authors and included in a final version of the document (FEIS). Supplemental issues are usually addressed in a supplemental EIS (SEIS).

FEDERAL NOISE REGULATIONS AND GUIDELINES

Agencies

Federal regulations and guidelines applicable to environmental noise assessment are those drafted by EPA, HUD, FHWA, FAA, FTA, FRA, OSHA, APTA, DoD, and GSA. The most recent versions of regulations and guidelines promulgated by these agencies are discussed below. It should be noted that the U.S. Environmental Protection Agency (EPA) had an Office of Noise Abatement and Control (established through the Noise Control Act of 1972, Public Law 92-574) that was closed by the Reagan Administration in 1982. At that point it was decided that the agencies listed below, along with state and local agencies, would regulate noise affairs. The EPA therefore does not currently deal with noise issues; however, there is a bill (the Office of Noise Abatement and Control Establishment Act of 1991) pending in the House of Representatives that, if enacted, would reinstate funding for the Office of Noise Abatement and Control.

Environmental Protection Agency

As sanctioned by the Noise Control Act of 1972, the EPA Office of Noise Abatement and Control (ONAC) published regulations between 1972 and 1982, under Subchapter G of Chapter I of Title 40, Code of Federal Regulations. The specific regulations are in Parts 201 through 211, their usual designation being 40 CFR Parts 201–211. 40 CFR Part 201 contains noise limits, with associated measurement procedures, for interstate rail (train) sources. 40 CFR Part 202 provides noise limits for motor vehicles involved in interstate commercial activities. 40 CFR Part 203 deals with certification procedures for "low-noise-emission products," as defined by the Noise Control Act of 1972. 40 CFR Part 204, although labeled to contain noise emission standards for construction equipment, deals principally with noise limits for portable air compressors with rated capacities of at least 75 ft^3/min and delivering air pressures of at least 50 pounds per square inch (psi). 40 CFR Part 205 contains noise limits, with associated measurement procedures, for trucks with gross weight ratings over 10,000 lb and motorcycles. 40 CFR Parts 209 and 210 deal with procedures in filing suits under the Noise Control Act of 1972. Section 12 of the Noise Control Act of 1972 was amended by the Quiet Communities Act of 1978 (Public Law 95-609) to provide civil penalties in addition to the original criminal sanctions. 40 CFR Part 210 describes procedures private citizens must use to enforce this act. 40 CFR Part 211 prescribes proper product labeling and testing of hearing protection devices (HPDs).

In 1981, Congress agreed with the White House Office of Management and Budget to cease funding for ONAC. Congress did not, however, repeal the Noise Control Act of 1972 when it made that decision. Because of this, the EPA is still legally responsible for enforcing 40 CFR Parts 201–211 without budget for their enforcement or updating. This has created a freeze on product labeling and noise emission standards because state and local governments are prohibited from adopting different standards (Shapiro, 1991).

Department of Housing and Urban Development

The U.S. Department of Housing and Urban Development (HUD) Noise Regulation was published on July 12, 1979, in the Federal Register as 24 CFR Part 51B. The regulation establishes standards for HUD-assisted projects and actions, requirements, and guidelines on noise abatement and control. The HUD regulations contain standards for exterior noise levels only. An upper limit L_{dn} goal of 45 dBA is set forth for interior noise levels and the attenuation requirements are geared toward achieving that goal, although interior standards are not contained in the regulation. Three categories of acceptability are established in the regulation and mitigation requirements for less than acceptable environments are given.

Federal Highway Adminstration

The U.S. Department of Transportation Federal Highway Administration (FHWA) Noise Regulation was published on July 8, 1982, in the Federal Register as 23 CFR Part 772. The regulation establishes standards for vehicular traffic noise studies on existing or proposed state or interstate highways. The FHWA divides projects into Types I and II categories. A Type I project is a proposed federal or federally aided project for the construction of a new highway or major physical alterations to an existing highway. A Type II project is a project for the construction of noise abatement measures that are added to an existing highway with no major alterations to the highway itself. The FHWA regulations apply to all Type I projects. Development and implementation of Type II projects are not mandatory, but a traffic noise analysis is required for eligibility of federal funding for abatement measures.

Five land use categories are listed in terms of not-to-exceed limits, denoted noise abatement criteria (NAC), along with qualitative relative impact criteria. The regulations are meant to be used and interpreted by individual state transportation agencies. Evaluation and mitigation guidelines are given. The FHWA noise abatement procedures are described in their *Federal-Aid Highway Program Manual* (FHPM), Volume 7, Chapter 7, Section 3, *Procedures for Abatement of Highway Traffic Noise and Construction Noise*, published August 9, 1982.

Federal Aviation Administration

The U.S. Department of Transportation Federal Aviation Administration (FAA) is the only agency of the federal government specifically directed by Congress to regulate aircraft for noise abatement procedures. Rules controlling aircraft noise are issued by the Administrator of the FAA after consideration of safety, economics, public health and welfare, and technology, and after consultation with the Secretary of Transportation, the EPA, and other interested agencies.

In response to NEPA, FAA Order 1050.1D (issued December 5, 1986), *Policies and Procedures for Considering Environmental Impacts*, contains provisions to ensure that the FAA air traffic regulatory actions are responsive to aircraft noise problems. Most FAA noise regulations promulgate standards for specific aircraft design noise limitations. In addition to the internal directive cited above, Federal Aviation Regulation (FAR) Part 150 (issued October 25, 1989), *Airport Noise Compatibility Planning*, establishes a standardized noise and land use compatibility program, including plotting noise contour maps around airports. This is in response to Title I of the Aviation Safety and Noise Abatement Act of 1979, Public Law 96-193.

In addition, 14 CFR Part 91, published September 25, 1991, requires a phasing out of all older, noisier aircraft flying over the continental United States by December 31, 1999. This was published in response to the Airport Noise and Capacity Act of 1990, passed by Congress on November 5, 1990. FAR Part 36, *Noise Standards: Aircraft Type and Airworthiness Certification*, as amended through December 22, 1988, divides aircraft into three so-called stages (denoted by the numbers 1, 2, and 3) according to their noise emissions (Stage 1 being the loudest aircraft and Stage 3 being the quietest). According to 14 CFR Part 91, only aircraft meeting Stage 3 requirements will be permitted after December 31, 1999.

On the international front, the International Civil Aviation Organization (ICAO) has standards similar to those of the FAA under its *International Standards and Recommended Practices—Environmental Protection*, Annex 16, Volume 1, 2nd ed. (issued November 1988) of their regulations. In addition to the aircraft noise emission standards, this volume includes atmospheric sound attenuation factors that agree with those given in SAE ARP 866A (listed in Table 3-2 of Chapter 3).

Federal Transit Adminstration

The U.S. Department of Transportation Federal Transit Administration (FTA) (before November 1991 known as the Urban Mass Transportation Administration [UMTA]), in its *Guidance Manual for Transit Noise and Vibration Impact Assessment* (Document UMTA-DC-08-9091-90-1, published in 1993), provides guideline criteria for mass transit transportation sources. The noise criteria for the onset of impacts vary according to land use categories and the predicted project noise levels. Impacts are determined on the basis of a sliding community response scale.

Federal Railroad Adminstration

The U.S. Department of Transportation Federal Railroad Administration (FRA) regulates interstate rail noise in accordance with the EPA noise standards published under 40 CFR Part 201. In addition, 49 CFR Part 210 contains regulations for the enforcement of the EPA regulations. These regulations became effective in January 1984.

Occupational Safety and Health Administration

The U.S. Department of Labor Occupational Safety and Health Administration (OSHA) Occupational Noise Exposure Standard was published on June 28, 1983, in the Federal Register as Part 1910, Section 1910.95 to Title 29 of the CFR (29 CFR 1910.95). The regulation establishes a standard for noise exposure in the workplace. This regulation includes the Hearing Conservation Amendment, known as such because it revised the noise standards passed in 1972 describing conditions and components for hearing conservation programs in the workplace. The original action on this began with the Walsh–Healy Act in 1969 (34 FR 7948) and the Occupational Safety and Health Act of 1970 (Public Law 91-596). Although these regulations are not used in community noise assessments, they can be used as a reference for rating methods and hazardous levels.

Agencies Providing Noise Guidelines

American Public Transit Association

The American Public Transit Association (APTA), in its *Guidelines for Design of Rapid Transit Facilities* (published in 1981), provides guidelines for rail transportation sources. All noise criteria are in terms of not-to-exceed limit goals for rail operations based on community noise acceptability criteria.

Department of Defense

The U.S. Department of Defense (DoD) has established Air Installation Compatible Use Zones (AICUZ) for noise generated by DoD aircraft operations. Department of Defense Instruction Manual 4165.57, issued November 8, 1977, defines these criteria for compatible land uses relative to accident potential zones (APZ) and noise zones. The three types of impacts evaluated by AICUZ studies are in terms of noise, accident potential, and obstruction height limitations. Specific procedures for implementing AICUZ programs are given in the *AICUZ Handbook* published by the U.S. Air Force (USAF, 1992).

Department of Veterans Affairs
The U.S. Department of Veterans Affairs (VA), in Section 2.52 of its Department of Veterans Benefits (DVB) Manual M26-2 (dated January 26, 1988), classifies airport noise zones for acceptability of security for VA-generated loans. These zones are similar to those designated by HUD for funding approval.

General Services Adminstration
The General Services Administration (GSA) specifies interior noise standards for public buildings in PBS PQ100, *Facilities Standards for the Public Building Service*, published September 1992. These standards are discussed in Chapter 7 of this book. As part of Section 01040 of the *GSA Supplement to Masterspec*, published May 1987, noise limits are placed on construction equipment used on GSA sites. These limits are discussed in this chapter.

Noise Assessment Methods

All methods used by Federal agencies for environmental noise assessments are discussed below. The discussion covers descriptors used, study area definition, and models and analysis techniques used.

Descriptors
The EPA recommends the use of the L_{dn} as the primary descriptor of environmental noise for land use compatibility planning (EPA, 1982). The EPA regulations in 40 CFR Parts 201-211 use maximum instantaneous dBA limits. Because the EPA is not currently involved in the environmental noise assessment process, it will be included below only in discussions where appropriate information can be rendered.

The descriptor required by HUD regulations is the 24-h L_{dn}. When dealing with HUD-assisted projects, *The Noise Guidebook* (HUD, 1985) provides methods to estimate vehicular traffic, rail, and aircraft noise in terms of L_{dn}.

The FHWA gives the state agency a choice of using either $L_{eq(1)}$ or $L_{10(1)}$, but not both, in its vehicular traffic noise regulation.

The FAA currently uses the annual average L_{dn} as its preferred noise descriptor. The annual, or yearly, L_{dn} is an average of daily L_{dn} values over the course of a year. The FAA has its own computer program to calculate L_{dn} contours based on measurements of typical flyover noise levels from individual aircraft. Many airports are also monitored to determine actual noise levels in their vicinities. Figures 5-1 through 5-4 show examples of L_{dn} contour maps for major international and medium-sized airports.

The FTA uses the sound exposure level (SEL) in conjunction with a choice of $L_{eq(1)}$ or L_{dn} as its principal noise descriptors for mass transit noise.

The OSHA uses the $L_{eq(8)}$ as its principal descriptor with a 5-dBA exchange rate. The exchange rate is the basis for doubling exposure levels. Typical L_{eq} calculations use a 3-dBA exchange rate, meaning that the noise exposure is doubled each time 3 dBA is added to the L_{eq} value. Years of OSHA studies have concluded that, in terms of potential noise-induced hearing loss, an addition of 5 dBA to the L_{eq} (above a certain threshold L_{eq}) would double the hazardous potential of the noise exposure such that half of the exposure time spent at the original L_{eq} would provide the same hazard at the higher L_{eq}. The 8-h time duration is used as a basis here because the OSHA deals with occupational hazards, which typically involve 8-h work shifts and, thus, 8-h exposures. The $L_{eq(8)}$ with a 5-dBA exchange rate is also known as the time-weighted average (TWA) value. Time-weighted average analysis extrapolates readings taken over periods of less than 8 h to 8-h exposure levels.

The APTA uses what it calls the single event maximum noise level as its principal rail noise descriptor. This translates to the maximum instantaneous dBA sound pressure level measured, using the fast meter response speed, during a train passby. The APTA noise guidelines, presented in their 1981 *Guidelines for Design of Rapid Transit Facilities*, state that train noise levels, because of their short duration, may appear acceptable on an energy-averaged basis but, because of the potentially large differences between maximum passby levels and the community ambient levels, train noise may be unacceptable because of its magnitude. Single event maximum noise levels have been chosen as the descriptor for this reason.

The DoD and VA criteria are in terms of L_{dn}. The GSA construction noise limits are in terms of not-to-exceed values at 50 ft.

Applicability of Descriptors to Environmental Conditions

Traffic noise is typically expressed in terms of $L_{eq(1)}$, although FHWA criteria were originally specified in terms of $L_{10(1)}$ only. The $L_{10(1)}$ descriptor is inadequate when hourly traffic flow rates are low, when vehicles are not evenly spaced on a road, and when attempts are made to mathematically combine $L_{10(1)}$ values with other descriptors. Because these disadvantages do not apply to $L_{eq(1)}$ values, it is gradually becoming the descriptor of choice in most states for vehicular traffic noise analysis. One disadvantage of the L_{eq} descriptor is that it is extremely sensitive to discrete high-level events. For this reason, a complete picture of the noise environment can be

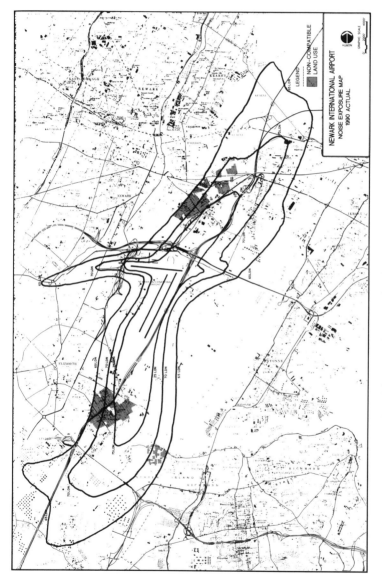

FIGURE 5-1. L_{dn} contours plotted for Newark International Airport, for illustration purposes only. The latest noise contours are probably different from these because of changing fleet mixes and operations. (Courtesy of The Port Authority of New York and New Jersey. With permission.)

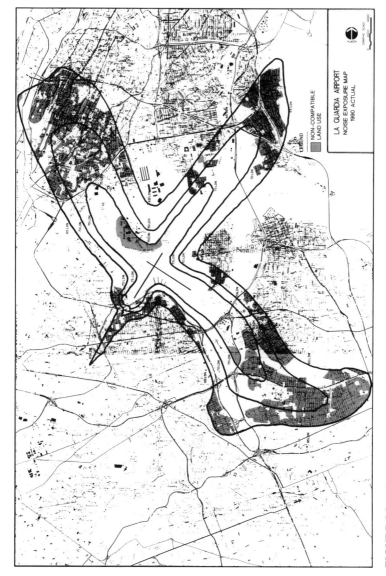

FIGURE 5-2. L_{dn} contours plotted for La Guardia Airport, for illustration purposes only. The latest noise contours are probably different from these because of changing fleet mixes and operations. (Courtesy of The Port Authority of New York and New Jersey. With permission.)

167

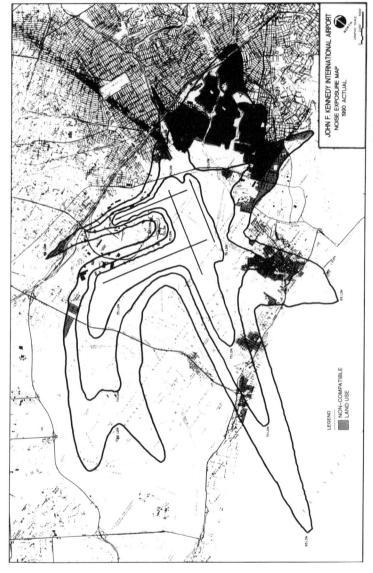

FIGURE 5-3. L_{dn} contours plotted for John F. Kennedy International Airport, for illustration purposes only. The latest noise contours are probably different from these because of changing fleet mixes and operations. (Courtesy of The Port Authority of New York and New Jersey. With permission.)

168

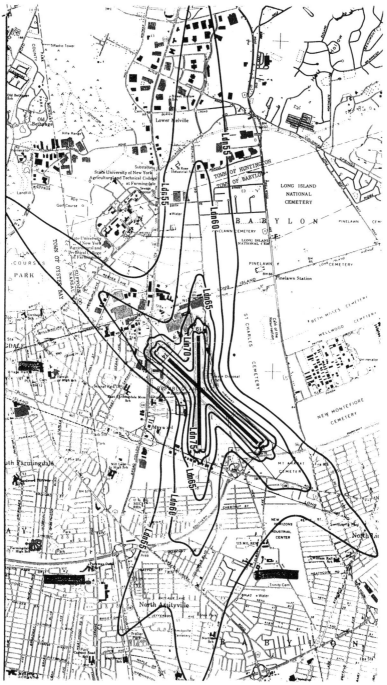

FIGURE 5-4. L_{dn} contours plotted for a medium-sized airport, for illustration purposes only. The latest noise contours are probably different from these because of changing fleet mixes and operations. (Courtesy of Republic Airport, East Farmingdale, NY. With permission.)

obtained by viewing the percentile levels as backup information for the L_{eq} value.

As mentioned in Chapter 2, a common criticism of energy-averaged rating methods such as L_{eq} or L_{dn} is that they smooth out discrete high-level events, such as aircraft flyovers or train passbys, to the point of eliminating the annoyance factor of the events. This is why rail noise guidelines have been changed recently from an L_{eq} basis to a single event maximum noise level basis. For the same reason, there has been much debate between the FAA and other organizations to change aircraft noise impact descriptors from L_{dn} to one that deals more realistically with the annoyance of each aircraft flyover. There is still agreement among most agencies, however, that L_{dn} is the best single descriptor to use for environmental noise assessments.

Identifying Receptor Locations

For all types of noise sources, noise receptor locations are typically at the closest property line or building location of a noise-sensitive location to the source in question (outdoors) or 3 ft from the closest window to the source in question (indoors). As stated in Chapter 3, the measuring microphone should be at least 3 ft away from any reflective surfaces to avoid contamination of the measurements. Any obstruction to the sound path can be thought of as having reflective qualities.

Study Area Identification

The U.S. Department of Housing and Urban Development clearly defines the noise assessment study area for different types of sources. *The Noise Guidebook* (HUD, 1985) specifies that, with respect to a proposed project site, all areas within 1000 ft of significant roadways, all areas within 3000 ft of rail operations, and all areas within 15 mi of military or civilian airports be included in the noise assessment study area.

The FHWA is much more general in its approach to study area. The FHWA regulations specify that activities and land uses that may be affected by traffic noise must be identified by field inspection, land use maps, and highway plans. After determining where human activity typically occurs, noise receptor locations may be chosen away from localized noise sources such as building air conditioners, barking dogs, or operating lawn mowers.

The study area for airports, according to the FAA, would encompass the area within L_{dn} 65 contours, where noise-sensitive locations exist.

The FTA provides a list of screening distances from noise-sensitive land uses to the nearest right-of-way or project site boundary. These are shown in Table 5-2, in terms of the type of project.

The OSHA considers the workplace only.

The APTA criteria in *Guidelines for Design of Rapid Transit Facilities* (APTA, 1981) divide communities into five categories and provide not-to-exceed limits for three different types of building use within each community area category. The study area is defined to begin at any point at least 50 ft from the rail track centerline. Different limits are provided for airborne and ground-borne noise caused by rail operations. Regardless of the community area, specific limits are placed on certain noise-sensitive buildings such as amphitheaters, quiet outdoor recreation areas, concert halls, radio and television studios, auditoriums, churches, theaters, schools, hospitals, museums, and libraries.

TABLE 5-2   **Federal Transit Administration Screening Distance for Noise Assessments**

Type of Project	Distance (ft)
Fixed guideway (rail/trolley) systems	
Commuter rail mainline	750
Commuter rail station	300
Rail transit guideway	750
Rail transit station	300
Access roads	100
Low- and intermediate-capacity transit	100
Steel wheel	750
Rubber tire	500
Yards and shops	2000
Parking facilities	250
Access roads	100
Ancillary facilities	
Ventilation shafts	200
Power substations	200
Bus systems	
Busway	750
Bus facilities	
Access roads	100
Transit mall	250
Transit center	450
Storage and maintenance	500
Park and ride lots	250

Source: Hanson et al. (1993).

The DoD and VA consider noise-sensitive areas within L_{dn} 65 contours. The GSA considers noise levels 50 ft from construction sources.

Models and Analysis Techniques

Department of Housing and Urban Development

The U.S. Department of Housing and Urban Development uses not-to-exceed L_{dn} values for its noise assessment. *The Noise Guidebook* (HUD, 1985) suggests predicting separate L_{dn} values for aircraft, vehicular, and rail noise sources and combining the three into a total L_{dn} to be used as the value to compare with the limits of the regulation. The screening method used for noise assessment is to determine whether any airports exist within 15 mi, any roadways exist within 1000 ft, or any rail lines (except totally covered subways) exist within 3000 ft of the proposed site. If none of these exist, no additional analysis is required. If any of these situations do exist, the following analysis must be performed.

For aircraft noise, if there is an airport within 15 mi of the site, current L_{dn} contours available from the FAA or calculated by FAA programs, or the charts provided in *The Noise Guidebook* (HUD, 1985), must be used. The charts given in *The Noise Guidebook* are in terms of the number of subsonic jet operations (takeoffs and landings) and the distances between the so-called noise assessment location (NAL) and the flight paths. The aircraft L_{dn} value is then interpolated off the L_{dn} maps given, or calculated. If L_{dn} data are not available but CNEL data are, the CNEL data can be used to approximate the L_{dn} data.

For roadway noise, if there is a road within 1000 ft of the site, currently available noise predictions or predictions calculated by the nomograph method in *The Noise Guidebook* must be used. The HUD prediction method takes into account the average daily traffic (ADT), the 24-h average of traffic volumes in both directions on a roadway, for automobiles and heavy trucks (weighing more than 26,000 lb with three or more axles). Buses capable of carrying more than 15 seated passengers are considered heavy trucks. Smaller buses are considered medium trucks. Medium trucks (weighing between 10,000 and 26,000 lb) are counted as equivalent to 10 automobiles. The HUD method also takes into account road gradient (if 2% or more), average traffic speed, the distance from the NAL to the near and far edges of the roadway lanes, the distances to stop signs, and the fraction of ADT that occurs during the nighttime hours of 10:00 P.M. to 7:00 A.M. If this fraction is unknown, 15% is used for both trucks and autos. Adjustments are given for all of these conditions in the calculations. Barrier attenuation nomographs are also given. After taking all of these factors into account and using all charts and nomographs, a single vehicular traffic L_{dn} value is derived.

If an FHWA highway prediction or noise measurement study is available in terms of the worst-case (loudest) hour $L_{eq(1)}$ or $L_{10(1)}$, the L_{dn} can be approximated by $L_{eq(1)}$ or $L_{10(1)} - 3$ for the loudest hour under the conditions that heavy trucks do not exceed 10% of the total traffic over 24 h and the traffic volume between 10:00 P.M. and 7:00 A.M. does not exceed 15% of the average daily traffic volume over 24 h.

For rail noise, if there is a railway within 3000 ft of the site, *The Noise Guidebook* (HUD, 1985) provides a noise prediction method with nomographs and adjustment factors similar to those used for vehicular traffic. The data taken into account are the distance from the NAL to the center of the railway track carrying the most traffic, the number of diesel and electric trains in both directions over an average 24-h day, the fraction that operate at night (10:00 P.M. to 7:00 A.M., assumed 15% if unknown), the average number of diesel locomotives per train (assumed 2 if unknown), the average number of railway cars per diesel or electric train (assumed 50 for diesel and 8 for electric if unknown), the average train speed (assumed 30 mph if unknown), if the track is made from welded or bolted rails, and the proximity of on-grade railroad crossings and whistle posts telling the engineer to blow the horn or whistle. After taking all of these factors into account, and using all charts and nomographs, a single railway L_{dn} value is derived.

The L_{dn} values corresponding to aircraft, vehicular traffic, and rail operations are then combined logarithmically to produce a single L_{dn} that is rated against the HUD noise assessment categories.

Federal Highway Administration

Whenever an FHWA Type I project is to be constructed, FHWA regulations require that existing noise levels at representative noise-sensitive locations be measured or calculated and that future noise levels with the proposed project are predicted, using a methodology acceptable to FHWA. If existing $L_{eq(1)}$ or $L_{10(1)}$ levels are measured, they should be performed during the period when highest noise levels would be generated and during the most appropriate hours for the affected land uses.

The FHWA has a traffic noise prediction model that can be used manually with nomographs and adjustment factors or with a computer in the form of the program STAMINA 2.0. STAMINA 2.0 is the prediction methodology most commonly used by the FHWA and most state transportation agencies. It may be used to analyze complex highway sites with many roadways, barriers (using the companion program OPTIMA), and receivers. Each roadway may be divided into straight line segments (called links) and roadways and barriers are analyzed in terms of their topographical orientation. Adjustments are automatically performed for different vehicle classifications (autos, medium trucks, heavy trucks, and five user-defined classes). The vehicle speed range is 30 to 65 mph. Any speeds below 30 mph are defaulted to 30 mph, and constant speed is assumed. As data have shown that SEL values for heavy trucks can vary by up to 12 dBA in acceleration and deceleration conditions, there are procedures that have been developed to account for such occurrences. Grade and barrier absorption are also taken into account in the analysis.

Running the STAMINA 2.0 program results in $L_{eq(1)}$ and $L_{10(1)}$ for each identified receiver link. The OPTIMA barrier design program can be used to calculate barrier performance and cost effectiveness evaluation. It is common practice to calibrate the STAMINA model results by comparing monitored to predicted noise levels at the same locations. An agreement between measured and predicted levels within ± 2 dBA is usually considered acceptable. If such agreement is not found but the differences are consistent (i.e., only positive or only negative and the predictions are within ± 2 dBA of each other), a correction factor can be added to the predicted levels. If the lack of agreement is not consistent (i.e., sometimes positive and sometimes negative), the prediction model may not be appropriate for that case and other models that calibrate acceptably would have to be used.

Once existing and future $L_{eq(1)}$ or $L_{10(1)}$ levels for the worst-case noise hour are measured and/or calculated, they are compared with the FHWA impact criteria.

Federal Aviation Administration

Hand calculation methods are available for single-event flyover noise prediction; however, they can be cumbersome. For environmental noise assessment, when L_{dn} contours of airports are required and not available from the FAA or airport personnel, the FAA Integrated Noise Model (INM) computer program is regarded as the standard prediction analysis tool for airport noise exposure modeling. Given the flight mix and number of daily operations, INM can be used to predict L_{dn} contours for fixed-wing sources and the Heliport Noise Model (HNM) can be used for helicopter sources. A simplified version of INM, known as the Area Equivalent Method (AEM), can be used as a screening analysis to determine the need for an EIS due to added aircraft operations. The AEM calculates the increase or decrease of the area covered by the L_{dn} 65 contour resulting from changes in aircraft operations and/or flight mix.

Federal Transit Administration

The FTA has established three levels of noise assessment for related projects. The first level is a screening procedure to identify locations where a proposed project has little possibility of noise impact. This is based on whether any noise-sensitive locations exist within the distances listed in Table 5-2. If no noise-sensitive locations are within those limits, no noise analysis is needed for the project.

The second level of assessment is the general assessment, used when the screening procedure identifies noise-sensitive locations and a broad approach is enough to evaluate impacts and propose mitigation measures. The procedure involves noise predictions based on charts of SEL data for potential sources, combined with correction factors. These factors, for rail operations, involve average number of vehicles per train, volume of trains per hour, average daytime (7:00 A.M. to 10:00 P.M. and nighttime (10:00 P.M. to 7:00 A.M.) volumes, maximum train speed, closest distance between track centerline and receptor, and noise barrier location. For highway operations (buses, vans, and car pools), the correction factors involve volumes per hour, average daytime and nighttime volumes, maximum speeds, and barrier locations. For stationary sources (e.g., terminals and parking facilities), correction factors involve average numbers of autos, buses, and trains per hour, and noise barrier locations.

Existing ambient noise estimates are obtained from calculations or charts, based on population density in people per square mile and distances from major roadways. A noise impact contour is then plotted on an area map based on charts of impact criteria and level vs. distance corrections.

The third level of assessment is the detailed analysis that is needed when site-specific mitigation requirements are necessary. At this level of analysis, measurements are preferred, but not required, for some vehicles and equipment. Measurements are required for existing ambient level, preferably full 1-h and 24-h (for residential receptors), but full values can be extrapolated from partial measurements. Measurement methods are deferred to accepted industry practice.

Calculations similar to those required for the general assessment are suggested for rail, highway, and stationary sources, with additional factors to be included. For rail sources, these are consideration of warning horns and the average power rating and throttle setting of the train locomotive. For highway sources, emission factors for three categories of buses are evaluated along with road surfaces for automobile travel. Stationary source analysis accounts for discrete events such as signal crossing alarms, buses idling, or car washes. The calculations also include a much more detailed barrier analysis than is given in the general assessment.

As for the general assessment, an L_{dn} or $L_{eq(1)}$ is derived for existing ambient and future project levels. For the detailed analysis, the total level for rail and bus facilities is plotted in contours similar to the general assessment type, with several lines to delineate different degrees of impact zones. Highway sources are evaluated according to FHWA impact criteria.

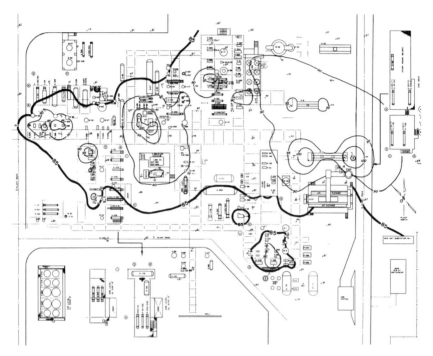

FIGURE 5-5.   dBA noise contours for a sample refinery.   (Courtesy of Conoco, Inc. With permission.)

All derived noise levels are rated against a sliding impact criterion curve, based on three categories of noise-sensitive land uses.

Occupational Safety and Health Administration

The OSHA regulations are in terms of measured $L_{eq(8)}$ values having a 5-dBA exchange rate, sometimes denoted L_{OSHA}. Typically, contours of constant SPL are mapped onto plant layout drawings, similar to elevation drawings of topographical maps. These contour maps are used to identify potentially hazardous areas. Typical noise contour maps are shown in Figs. 5-5 through 5-7. Employees representing different job assignments are also required to wear noise dosimeters to monitor employee noise exposure levels. A dosimeter (or dose meter) is a small box, usually the size of a small calculator, that can be strapped through a belt around an employee's waist, with a cord and a small microphone that is fastened by a clip to the employee's shirt collar so that the microphone is as close as possible to the employee's ear. A sample dosimeter is shown in Fig. 5-8.

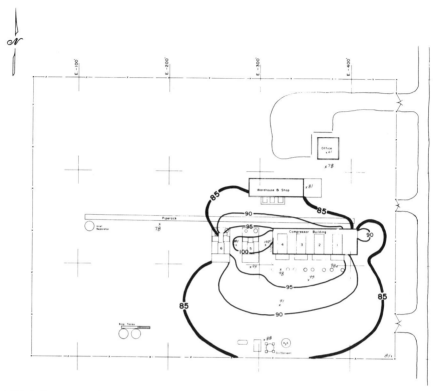

FIGURE 5-6. dBA noise contours for a sample compressor station. (Courtesy of Conoco, Inc. With permission.)

The value read by the dosimeter is the time-weighted average (TWA). As mentioned earlier, this value takes the noise reading over a certain period of time, assuming that the person will be in that type of environment for an 8-h period. For example, if a worker were exposed to 90 dBA of TWA noise for 8 h, the dosimeter would read 90 dBA for the exposure. If the same worker were exposed to 90 dBA for 4 h and the dosimeter was turned off at that point, the dosimeter would assume that the worker would be exposed to 90 dBA for the entire 8-h period and read the same 90-dBA level.

American Public Transit Association

The criteria from *Guidelines for Design of Rapid Transit Facilities* (APTA, 1981) are in terms of design goals and absolute noise limits for rail operations. No prediction or screening methodologies are mentioned in the guidelines. Measurement procedures are discussed that suggest monitoring in free field

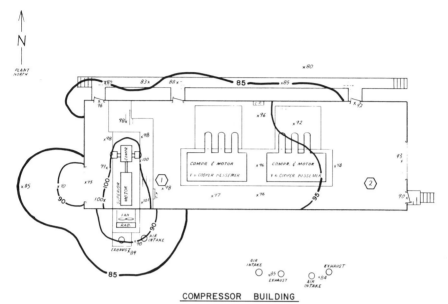

FIGURE 5-7.   dBA noise contours for a sample compressor building.   (Courtesy of Conoco, Inc. With permission.)

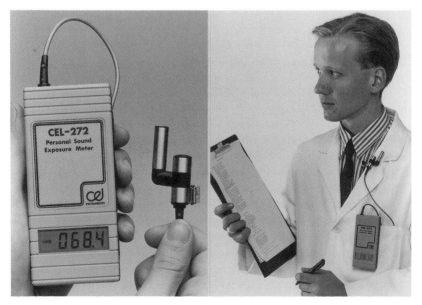

FIGURE 5-8.   A noise dosimeter.   (Courtesy of Lucas Industrial Instruments, Severna Park, MD. With permission.)

environments where no reflecting or shielding surfaces exist. The SPL is to be monitored in dBA, using a sound level meter that meets ANSI type 2 specifications. The slow meter response speed is to be used for all measurements except those involving moving and transient sources such as exterior train noise, train noise from vent shafts, or from door movements. These sources should be monitored by the fast meter response speed.

Department of Defense
The *AICUZ Handbook* (USAF, 1992) discusses the detailed analysis program that is involved in analyzing noise around military airports. The methods involve using several computer programs for data gathering and processing. The final output goes through the NOISEMAP program to produce L_{dn} contours, similar to the INM model that the FAA uses. The AICUZ study can be used as part of NEPA process documents (i.e., FONSIs, EAs, and EISs).

Veterans Administration
The DVB Manual M26-2 (VA, 1988) classifies L_{dn} noise levels in terms of zones and does not provide any analysis recommendations.

General Services Administration
The GSA provides construction equipment noise limits at 50 ft with no analysis recommendations except to comply with all applicable state and local laws, ordinances, and regulations relating to noise control.

Acceptable Levels
Acceptable noise levels are discussed below as provided by the EPA, HUD, FHWA, FAA, FTA, APTA, DoD, VA, and OSHA.

Environmental Protection Agency
In response to the Noise Control Act of 1972, the EPA published *Information on Levels of Environmental Noise Requisite to Protect Public Health and Welfare with an Adequate Margin of Safety* in 1974. The Levels Document, as it is commonly known, summarized acceptable noise limits as is shown in Table 5-3.

The 70-dBA hearing loss limit is based on 40 years of exposure time.

Department of Housing and Urban Development
The HUD regulations consider outdoor L_{dn} noise levels of 65 dBA or less to be acceptable for residential developments. It is assumed that standard building construction would provide a minimum of 20 dBA of noise attenuation and, therefore, an interior L_{dn} limit of 45 dBA is the goal of acceptable

TABLE 5-3   Noise Levels Identified as Requisite to Protect Public Health and Welfare with an Adequate Margin of Safety

Effect	Level (dBA)	Area
Hearing loss	$L_{eq(24)} \leq 70$	All areas
Outdoor activity interference and annoyance	$L_{dn} \leq 55$	Outdoors in residential areas and farms and other outdoor areas where people spend widely varying amounts of time, and other places where quiet is a basis for use
	$L_{eq(24)} \leq 55$	Outdoor areas where people spend limited amounts of time, such as school yards, playgrounds, etc.
Indoor activity interference and annoyance	$L_{dn} \leq 45$	Indoor residential areas
	$L_{eq(24)} \leq 45$	Other indoor areas with human activities such as in schools, etc.

Source: EPA (1974).

levels. Attenuation requirements in the regulations are geared toward this interior goal, although no interior noise standards are promulgated in the regulations.

Table 5-4 shows the HUD's complete land use compatibility guidelines.

Federal Highway Administration

The FHWA regulations consider a maximum indoor $L_{eq(1)}$ of 52 dBA or $L_{10(1)}$ of 55 dBA and a maximum outdoor $L_{eq(1)}$ of 57 dBA or $L_{10(1)}$ of 60 dBA to constitute an acceptable environment for any land use. These maximum outdoor limits are meant for extremely sensitive areas. Typical residential areas have limits that are 10 dBA higher.

Federal Aviation Administration

The FAA considers an outdoor L_{dn} of less than 65 dBA to constitute an acceptable environment. No interior levels are considered. The FAA land use compatibility guidelines are listed in Federal Mitigation Requirements (below) because they include guidelines that are appropriate for the mitigation discussion. There has been some discussion between the FAA and civic groups about lowering this limit; however, there is general agreement that prediction methods lose considerable accuracy and noise effects have not been established below L_{dn} 65 (FICON, 1992).

TABLE 5-4   U.S. Department of Housing and Urban Development Land Use Compatibility Guidelines[a]

Land Use Category	Clearly Acceptable	Normally Acceptable	Normally Unacceptable	Clearly Unacceptable
Residential, classrooms, churches, libraries, hospitals, nursing homes, sports arenas (indoor and outdoor)	< 60	60–65	65–75	> 75
Transient lodging	< 65	65–70	70–80	> 80
Auditoriums, concert halls, music shells	< 50	50–60	60–70	> 70
Playgrounds, parks	< 55	55–65	65–75	> 75
Golf courses, riding stables, water recreation facilities, cemeteries	< 60	60–70	70–80	> 80
Office buildings, retail, movie theaters, restaurants	< 65	65–75	75–80	> 80
Wholesale, industrial, manufacturing, utilities	< 70	70–80	80–85	> 85
Manufacturing, communication (sensitive)	< 55	55–70	70–80	> 80
Livestock farming, animal breeding	< 60	60–75	75–80	> 80
Agriculture (except livestock), mining, fishing	< 75	75–95	No rating	No rating
Public right-of-way	< 75	75–85	85–95	No rating
Extensive natural recreation areas	< 60	60–75	75–85	> 85

[a] In dBA L_{dn}.

Source: HUD (1985).

Federal Transit Administration

The FTA *Guidance Manual for Transit Noise and Vibration Impact Assessment* (Hanson et al., 1993) deals with the acceptability of noise levels in terms of comparing project-generated levels with existing ambient L_{dn} or $L_{eq(1)}$ levels. As a lower limit, L_{dn} and $L_{eq(1)}$ levels below 52 dBA are generally considered acceptable in any environment. As ambient levels increase above 45 dBA, these acceptable level limits rise to a maximum of 70 dBA, as is shown in Impact Criteria (below). The L_{dn} is used when nighttime activity must be considered and $L_{eq(1)}$ is considered when only daytime operations would exist.

Occupational Safety and Health Administration

The OSHA considers occupational areas with $L_{eq(8)}$ less than 85 dBA to be non-hazardous in terms of their potential for causing noise-induced hearing loss.

American Public Transit Association

The criteria from *Guidelines for Design of Rapid Transit Facilities* (APTA, 1981) are in terms of not-to-exceed instantaneous noise levels. These limits are based on typical L_{50} data quoted for five different categories of land uses. These values are shown in Table 5-5.

Department of Defense and Veterans Administration

The DoD and VA consider outdoor areas where the L_{dn} is less than 65 dBA to constitute acceptable environments. The *AICUZ Handbook* (USAF, 1992) provides detailed land use compatibility guidelines for military aircraft facilities. These are provided in generalized form in Table 5-6. *Yes* means that the land use is compatible and *No* means that the land use in not compatible with the noise levels.

Impact Criteria

Environmental Protection Agency

The EPA regulations provide the following noise limits for the sources referenced. Noise levels from any of these sources should not be recorded if the maximum source levels, using the fast meter response speed, are less than

TABLE 5-5   American Public Transit Association Community Categories along Transit System Corridors

Category	Description	Typical Ambient Noise Level (dBA)
I	Low-density urban residential, open space park, suburban	40–50 (day) 35–45 (night)
II	Average urban residential, quiet apartment and hotels, open space, suburban residential, or occupied outdoor area near busy streets	45–55 (day) 40–50 (night)
III	High-density urban residential, average semiresidential/commercial areas, parks, museum and noncommercial public building areas	50–60 (day) 45–55 (night)
IV	Commercial areas with office buildings, retail stores, etc., primarily daytime occupancy; central business district	50–70
V	Industrial areas or freeway and highway corridors	Over 60

Source: APTA (1981).

TABLE 5-6   Department of Defense Aircraft Noise Land Use Compatibility Guidelines[a]

	Noise Contours			
Land Use Category	65–70	70–75	75–80	80–85
Residential	No[b]	No[b]	No	No
Public assembly, public services	No	No	No	No
Recreation	Yes	Yes	No	No
Transportation, communications, utilities, offices, commercial	Yes	Yes	Yes	No
Manufacturing, agriculture, mining	Yes	Yes	Yes	Yes

[a] In dBA L_{dn}.
[b] Unless sound attentuation materials are installed.

Source: USAF (1992).

10 dBA above the background levels without the source in question. Beside this, all other measurement procedures prescribed conform with those mentioned in Chapter 3. Warning devices, such as horns, are not included in these regulations.

Rail (40 CFR Part 201).   Stationary locomotive noise levels shall not exceed 87 dBA at 100 ft from the geometric center at any throttle setting except idle. While in idle, levels shall not exceed 70 dBA under the same conditions. These levels should be measured with the slow meter response speed. For moving locomotives, noise levels at 100 ft from the track centerline shall not exceed 90 dBA, using the fast meter response speed. This assumes that the track has less than a 2° curve or a radius of curvature greater than 2865 ft. For moving rail cars, noise levels should not exceed, at 100 ft from the track centerline as defined above, 88 dBA at speeds up to and including 45 mph and 93 dBA at speeds above 45 mph, using the fast meter response speed. Noise levels from retarders should not exceed 83 dBA, using the fast meter response speed, at any property line not owned or operated by the railroad. Car coupling operations shall not exceed 92 dBA, using the fast meter response speed, at any property line not owned or operated by the railroad.

In terms of locomotive load cell test stands, noise levels shall not exceed 78 dBA, using the slow meter response speed, at 100 ft from the geometric center of the locomotive. All test cells are in compliance with this regulation if their associated noise levels do not exceed 65 dBA at any property line not owned or operated by the railroad. If these limits cannot be met, levels must not exceed 65 dBA, using the fast meter response speed, at any property lines not owned or operated by the railroad at least 400 ft away from the geometric center of the locomotive being tested.

Interstate Motor Vehicles (40 CFR Part 202).   Under stationary conditions, no motor carrier shall operate a vehicle such that noise levels exceed 85 dBA, using the fast meter response speed, at 50 ft from the vehicle centerline when the engine is accelerated from idle to the governed speed, with the transmission in neutral and the clutch engaged. Under moving conditions, the noise limits at 50 ft from the lane of travel centerline, using the fast meter response speed, are 83 dBA for speed limits of 35 mph or less and 87 dBA for speed limits of more than 35 mph.

Portable air compressors (40 CFR Part 204).   Portable air compressors shall not produce noise levels exceeding 76 dBA, based on an averaging of five measurements, each recorded 23 ft from each side and the top of the unit. Each side measurement should be 5 ft off the ground.

Medium and Heavy Trucks (40 CFR Part 205).   Maximum noise levels, using the fast meter response speed, should not exceed 83 dBA for vehicles manufactured after January 1, 1979 and 80 dBA for vehicles manufactured after January 1, 1988, using the prescribed measuring procedure. The noise monitoring should take place on a controlled test facility with the vehicle driving by the microphone at distances varying from 50 to 70 ft, as is shown in Fig. 5-9. Because road and environmental conditions vary considerably from place to place, a casual roadside reading exceeding the levels stated above would not necessarily mean that the vehicle in question is violating this regulation.

Motorcycles (40 CFR Part 205).   Maximum noise levels are monitored in a highly controlled test facility similar to that used for trucks mentioned above. The required general layout of the testing facility is similar to that shown in Fig. 5-9. The noise limits for typical street motorcycles beginning

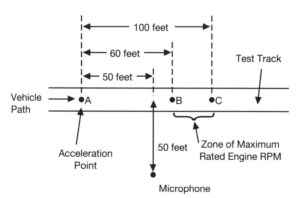

FIGURE 5-9.   Test facility setup for EPA vehicle noise monitoring.   (Adapted from 40 CFR Part 205.)

with model year 1983 are 83 dBA and beginning with model year 1986 are 80 dBA at the 50- to 70-ft range monitored on the course. The noise limit for moped-type motorcycles that cannot exceed 30 mph on a level paved surface is 70 dBA for model years beginning 1983. Off-road motorcycles with engine displacements of 170 cc or less, beginning with model year 1983, must not exceed 83 dBA and beginning in model year 1986 must not exceed 80 dBA. Off-road motorcycles with engine displacements greater than 170 cc, beginning with model year 1983, must not exceed 86 dBA and beginning in model year 1986 must not exceed 82 dBA. These limits are to be designed into motorcycles such that they will be met, with proper maintenance, for up to a year or 3730 mi for a street motorcycle or 1865 mi for an off-road motorcycle (whichever comes first), after they are sold.

For the purpose of low-noise-emission product (LNEP) certification, per 40 CFR Part 203, motorcycles bought by the federal government must meet more stringent noise limits. For motorcycles with engine displacements greater than 170 cc, the noise limit is 71 dBA for street vehicles and 75 dBA for off-road vehicles. For street and off-road motorcycles with engine displacements of 170 cc and lower, the noise limit is 71 dBA. For moped-type street motorcycles, the limit is 60 dBA. All of these levels correspond to maximum levels on the standard test track monitored at distances of 50 to 70 ft from the sources.

Department of Housing and Urban Development
The HUD regulations set forth the following exterior noise standards for new housing construction assisted or supported by the department:

$$\text{Acceptable: } L_{dn} \leq 65 \text{ dBA}$$

$$\text{Normally unacceptable: } 65 \text{ dBA} < L_{dn} \leq 75 \text{ dBA}$$

$$\text{Unacceptable: } L_{dn} > 75 \text{ dBA}$$

In the normally unacceptable range, 5 dBA of attenuation above that of standard construction must be provided in the L_{dn} 65 to 70 area and 10 dBA of additional attenuation must be provided in the L_{dn} 70 to 75 area. The HUD regulations provide only absolute, and no relative, impact criteria. It is assumed that standard construction provides 20 dBA of noise attenuation to interior spaces.

Federal Highway Administration
The FHWA identifies both relative and absolute impact criteria for vehicular traffic noise impacts. The relative criterion consists of a comparison between future predicted noise levels with the proposed project and existing levels. This type of impact is said to occur when the predicted future levels

TABLE 5-7   Federal Highway Administration Noise Abatement Criteria

Category	$L_{eq(1)}$ (dBA)	$L_{10(1)}$ (dBA)	Area Description
A	57 (exterior)	60 (exterior)	Lands on which serenity and quiet are of extraordinary significance and serve an important public need and where the preservation of those qualities is essential if the area is to continue to serve its intended purpose
B	67 (exterior)	70 (exterior)	Picnic areas, recreation areas, playgrounds, active sports areas, parks, residences, motels, hotels, schools, churches, libraries, and hospitals
C	72 (exterior)	75 (exterior)	Developed lands, properties, or activities not not included in categories A or B
D	—	—	Undeveloped lands
E	52 (interior)	55 (interior)	Residences, motels, hotels, public meeting rooms, schools, churches, libraries, hospitals, and auditoriums

Source: 23 CFR Part 772.

"substantially exceed" existing levels. This "substantial" term is left to individual state transportation agencies to define.

The absolute noise impact criterion is that future noise levels with the proposed project would not "approach or exceed" the noise abatement criteria (NAC) shown in Table 5-7. The "approach or exceed" phrase is also left to state transportation agencies to interpret and define.

In evaluating these NAC criteria, either the $L_{eq(1)}$ or $L_{10(1)}$ level, but not both, is to be used. These criteria identify impacts that are considered severe enough to warrant mitigation.

Federal Aviation Administration

The FAA has a relative threshold of significance identified as an increase of $1.5L_{dn}$ within the L_{dn} 65 contour at noise-sensitive locations. Recall that the FAA uses annual average L_{dn} (long term) values, so changes of only 1.5 dBA can still have significance in this case.

Federal Transit Administration

The FTA *Guidance Manual for Transit Noise and Vibration Impact Assessment* (Hanson et al., 1993) sets forth absolute noise impact criteria. The FTA has established three degrees of noise impact: no impact, impact, and severe impact. The choice of L_{dn} or worst-hour $L_{eq(1)}$ is given; the L_{dn} should be used for land uses where nighttime sensitivity is an issue and the L_{eq} should be used for land uses involving only daytime activities. Three categories of land use are employed, designated as categories 1, 2, and 3. Category 1 includes buildings and parks where quiet is an essential element of their intended purpose, such as outdoor concert pavilions. Category 2 includes residences and buildings where people normally sleep, such as hospitals and hotels. Category 3 includes institutional uses with primarily daytime and evening use, such as schools, libraries, churches, medical offices, concert halls, monuments, and museums. Commercial and industrial uses are not considered. It is recommended that the $L_{eq(1)}$ be used for categories 1 and 3 during the hours of sensitivity and the L_{dn} be used for category 2. This is because category 2 uses are the only ones with nighttime sensitivities. Table 5-8 shows the impact criteria.

Severe impacts would require mitigation whereas impacts would be analyzed in terms of the relative increase between existing and future Build levels and the types of noise-sensitive locations affected by the project.

Occupational Safety and Health Administration

The SPL exposure limits for occupational facilities are shown in Table 5-9. These limits are in terms of continuous, nonimpulsive noise. In addition to these prescribed limits, peak impulsive or impact SPLs are limited to 140 dB (unweighted).

The time-weighted average (TWA) calculation of the noise exposure compensates for different exposures and corresponding time durations, as listed in Table 5-9. It must be noted that the 90-dBA 8-h limit is lowered to 85 dBA when the exposed employee already has a standard threshold shift (STS), meaning that the employee has lost an average of 10 dB of hearing ability at 2000, 3000, and 4000 Hz in either ear. A formula has been established to determine whether a person is being overexposed to occupational noise, based on Table 5-9:

$$C_1/T_1 + C_2/T_2 + \cdots + C_n/T_n \leq 1 \qquad (5\text{-}2)$$

where C_n ($n = 1, 2, \ldots$) is the total daily exposure time to a specific noise level and T_n ($n = 1, 2, \ldots$) is the maximum permissible exposure time (according to Table 5-9) at that same noise level.

For example, consider an employee who is exposed to 90 dBA for 4 h, 100 dBA for 1 h, and 80 dBA for 3 h each day on the job. Using Eq. (5-2) in conjunction with Table 5-9, we would have the following:

$$C_1 = 4 \qquad T_1 = 8 \qquad C_2 = 1 \qquad T_2 = 2 \qquad C_3 = 3 \qquad T_3 = \infty \quad \text{(infinity)}$$

$T_3 = \infty$ because, according to the OSHA formula, anyone exposed to noise

TABLE 5-8 Federal Transit Administration Noise Impact Criteria[a]

Existing Ambient Level	Project Impact Levels (L_{dn} or $L_{eq(1)}$)					
	Category 1 or 2 Sites			Category 3 Sites		
(L_{dn} or $L_{eq(1)}$)	No Impact <Ambient + 10	Impact Ambient + 10–15	Severe Impact >Ambient + 15	No Impact <Ambient + 15	Impact Ambient + 15–20	Severe Impact >Ambient + 20
<43	<52	52–58	>58	<57	57–63	>63
43	<52	52–58	>58	<57	57–63	>63
44	<52	52–58	>58	<57	57–63	>63
45	<52	52–58	>58	<57	57–63	>63
46	<53	53–59	>59	<58	58–64	>64
47	<53	53–59	>59	<58	58–64	>64
48	<53	53–59	>59	<58	58–64	>64
49	<54	54–59	>59	<59	59–64	>64
50	<54	54–59	>59	<59	59–64	>64
51	<54	54–60	>60	<59	59–65	>65
52	<55	55–60	>60	<60	60–65	>65
53	<55	55–60	>60	<60	60–65	>65
54	<55	55–61	>61	<60	60–66	>66
55	<56	56–61	>61	<61	61–66	>66
56	<56	56–62	>62	<61	61–67	>67
57	<57	57–62	>62	<62	62–67	>67

58	<57	57–62	>62	<62	62–67	>67
59	<58	58–63	>63	<63	63–68	>68
60	<58	58–63	>63	<63	63–68	>68
61	<59	59–64	>64	<64	64–69	>69
62	<59	59–64	>64	<64	64–69	>69
63	<60	60–65	>65	<65	65–70	>70
64	<61	61–65	>65	<66	66–70	>70
65	<61	61–66	>66	<66	66–71	>71
66	<62	62–67	>67	<67	67–72	>72
67	<63	63–67	>67	<68	68–72	>72
68	<63	63–68	>68	<68	68–73	>73
69	<64	64–69	>69	<69	69–74	>74
70	<65	65–69	>69	<70	70–74	>74
71	<66	66–70	>70	<71	71–75	>75
72	<66	66–71	>71	<71	71–76	>76
73	<66	66–71	>71	<71	71–76	>76
74	<66	66–72	>72	<71	71–77	>77
75	<66	66–73	>73	<71	71–78	>78
76	<66	66–74	>74	<71	71–79	>79
77	<66	66–74	>74	<71	71–79	>79
>77	<66	66–75	>75	<71	71–80	>80

[a] In dBA.

Source: Hanson et al. (1993).

TABLE 5-9   Occupational Safety and Health Administration Permissible Noise Exposure Limits

Time Duration per day (h)	SPL Exposure Level (dBA)
8	90
6	92
4	95
3	97
2	100
1.5	102
1	105
0.5	110
0.25 or less	115

Source: 29 CFR Part 1910.95.

levels below 85 dBA can do so for any amount of time without risk of noise-induced hearing loss. Whenever ∞ is in the denominator of a fraction, the value of the fraction becomes 0. Therefore, whenever noise levels are below 85 dBA, their exposure times need not be included in Eq. (5-2). When dosimetry is performed, however, all noise levels between 80 and 130 dBA are to be included in the dosage calculations.

Continuing with this example, Eq. (5-2) would become:

$$4/8 + 1/2 = 1$$

Because the impact criterion is that the total be less than or equal to 1, this person would not be overexposed to noise as long as he does not have an STS. This value of 1 corresponds to a 100% exposure dose or a TWA of 90 dBA. A 50% exposure dose, or an 85-dBA TWA (based on the 5-dBA exchange rate), corresponds to the action level for hearing conservation programs, as discussed in Federal Mitigation Requirements (below).

American Public Transit Association

The criteria from *Guidelines for Design of Rapid Transit Facilities* (APTA, 1981) set forth instantaneous maximum noise level limits for airborne noise according to Tables 5-10 through 5-12.

Transient noise corresponds to events having short time durations, such as passby noises. Continuous noise corresponds to sources such as fans, cooling towers, or other sources producing noise of long duration. Transformer noise limits are 5 dBA less than those in Table 5-12.

TABLE 5-10   American Public Transit Association Airborne Noise Limits[a] for Train Operations

| | Single-Event Maximum Noise Level | | |
Community Category	Single Family Dwelling	Multifamily Dwelling	Commercial Building
I. Low-density residential	70	75	80
II. Average residential	75	75	80
III. High-density residential	75	80	85
IV. Commercial	80	80	85
V. Industrial/highway	80	85	85

[a] dBA at 50 ft from track center.

Source: APTA (1981).

TABLE 5-11   American Public Transit Association Airborne Noise Limits for Train Operations at Sensitive Locations

Building/Occupancy Type	Single-Event Maximum Noise Level (dBA)
Amphitheaters	60
Quiet outdoor recreation areas	65
Concert halls, radio and TV studios, auditoriums	70
Churches, theaters, schools, hospitals, museums, libraries	75

Source: APTA (1981).

TABLE 5-12   American Public Transit Association Noise Limits[a] from Transit System Ancillary Facilities

| | Maximum Noise Level | |
Community Category	Transient Noises	Continuous Noises
I. Low-density residential	50	40
II. Average residential	55	45
III. High-density residential	60	50
IV. Commercial	65	55
V. Industrial/highway	75	65

[a] dBA at 50 ft.

Source: APTA (1981).

Department of Defense

The AICUZ impact criteria for new military airfields are in terms of the land use compatibility criteria listed in Table 5-6. In addition, a change in L_{dn} of 2 dBA or more, resulting from any new operations at an existing airfield, at any noise-sensitive location, defines a significant change in noise exposure requiring a new AICUZ study.

Veterans Administration

Veterans Administration airport noise zones are defined in Table 5-13.

Proposed or existing properties in zone 1 are considered generally acceptable as security for VA-guaranteed loans. Proposed construction to be located in zone 2 would be acceptable if noise attenuation measures were built into the dwelling to ensure that interior noise levels are L_{dn} 45 or less. Proposed construction in zone 3 is considered generally not acceptable. Existing dwellings in zones 2 and 3 will not be rejected if the buyer is fully aware of the noise environment and the dwelling is equipped with year-round air conditioning and completely insulated.

General Services Administration

The GSA limits on construction noise sources are listed in Table 5-14.

Federal Mitigation Requirements

Department of Housing and Urban Development

The Department of Housing and Urban Development provides three alternatives for mitigating high noise exposures to projects assisted by HUD. The first suggestion is to relocate the noise-sensitive location. For example, if a housing development is proposed to be located near a highway where exterior noise levels would be above the acceptable range, the buildings may be set back and separated from the highway by a parking lot or park.

If relocation is not feasible, barriers are suggested. These barriers could be walls, berms, or other buildings. Erection of barriers would be effective only on vehicular traffic and nonelevated rail sources. The final alternative suggested is to provide noise attenuation to the interiors of buildings, using noise control design on building exteriors. These measures would include sealing windows, doors, and vents and, because typical windows and doors do not provide more than 25 dBA of attenuation, installing windows and doors that are designed to provide the required attenuation. As is shown in Chapter 4, double-paned and laminated windows are available that provide up to 40 dBA of attenuation. This method is considered least desirable because it leaves outdoor areas exposed to high noise levels.

TABLE 5-13   Veterans Administration
              Airport Noise Zones

Noise Zone	L_{dn} (dBA)
1	Under 65
2	65–75
3	Over 75

Source: VA (1988).

Federal Highway Administration

The FHWA states that where noise impacts are identified, state transportation agencies must examine ways to reduce substantially or eliminate the impact. Specifically, state transportation agencies must consider the opinions of the impacted residents, identify the abatement measures to be incorporated

TABLE 5-14   General Services Admini-
              stration Construction
              Noise Limits

Equipment	dBA (at 50 ft)
Front loader	79
Backhoe	85
Dozer	80
Tractor	80
Scraper	88
Grader	85
Truck	91
Paver	89
Concrete mixer	85
Concrete pump	82
Crane	83
Derrick	88
Pump	76
Generator	78
Compressor	81
Pile driver	101
Jack hammer	88
Rock drill	98
Pneumatic tool	86
Saw	78
Vibrator	76

Source: GSA (1987).

into the project as well as the impacts where no solutions are feasible, and include abatement measures in the approved project plans and specifications.

Noise abatement measures to be considered include barrier construction, road alterations, acquisition of land for buffer zones, traffic management strategies (e.g., prohibiting trucks, restricting hours of use, or modifying speed limits), and noise attenuation measures in impacted building exteriors.

TABLE 5-15   Federal Aviation Administration Land Use Compatibility with Yearly L_{dn} levels[a]

Land Use	L_{dn} (dBA)					
	<65	65–70	70–75	75–80	80–85	>85
Residential						
Residential, other than mobile homes and transient lodgings	Y	N (1)	N (1)	N	N	N
Mobile home parks	Y	N	N	N	N	N
Transient lodgings	Y	N (1)	N (1)	N (1)	N	N
Public use						
Schools	Y	N (1)	N (1)	N	N	N
Hospitals and nursing homes	Y	25	30	N	N	N
Churches, auditoriums, concert halls	Y	25	30	N	N	N
Government services	Y	Y	25	30	N	N
Transportation	Y	Y	Y (2)	Y (3)	Y (4)	Y (4)
Parking	Y	Y	Y (2)	Y (3)	Y (4)	N
Commercial use						
Offices, business/professional	Y	Y	25	30	N	N
Wholesale and retail building materials, hardware, and farm equipment	Y	Y	Y (2)	Y (3)	Y (4)	N
Retail trade—general	Y	Y	25	30	N	N
Utilities	Y	Y	Y (2)	Y (3)	Y (4)	N
Communication	Y	Y	25	30	N	N
Manufacturing and production						
Manufacturing—general	Y	Y	Y (2)	Y (3)	Y (4)	N
Photographic and optical	Y	Y	25	30	N	N
Agriculture (except livestock) and forestry	Y	Y (6)	Y (7)	Y (8)	Y (8)	Y (8)
Livestock farming and breeding	Y	Y (6)	Y (7)	N	N	N
Mining and fishing, resource production and extraction	Y	Y	Y	Y	Y	Y

(continued)

TABLE 5-15   (continued)

Land Use	L_{dn} (dBA)					
	< 65	65–70	70–75	75–80	80–85	> 85
Recreational						
Outdoor sports arenas and spectator sports	Y	Y (5)	Y (5)	N	N	N
Outdoor music shells, amphitheaters	Y	N	N	N	N	N
Nature exhibits and zoos	Y	Y	N	N	N	N
Amusement parks, resorts, camps	Y	Y	Y	N	N	N
Golf courses, riding stables, water recreation	Y	Y	25	30	N	N

[a] Key: Y, Land use and related structures are compatible without restrictions; N, land use and related structures are not compatible and should be prohibited; 25 or 30, land use and related structures are generally compatible; however, measures to achieve 25 or 30 dBA of attenuation must be incorporated into the design and construction of the structure. *Notes*: (1) Where the community determines that residential or school uses must be allowed, measures to achieve outdoor to indoor noise level reduction (NLR) of at least 25 and 30 dBA should be incorporated into building codes and be considered in individual approvals. Normal residential construction can be expected to provide an NLR of 20 dBA. The requirements normally assume alternate ventilaton and closed windows year round. The NLR criteria will not eliminate outdoor noise problems; (2) measures to achieve an NLR of 25 dBA must be incorporated into the design and construction of portions of these buildings where the public is received, office areas, noise-sensitive areas, or where the normal noise level is low; (3) measures to achieve an NLR of 30 dBA must be incorporated into the design and construction of portions of these buildings where the public is received, office areas, noise-sensitive areas, or where the normal noise level is low; (4) measures to achieve an NLR of 35 dBA must be incorporated into the design and construction of portions of these buildings where the public is received, office areas, noise-sensitive areas, or where the normal noise level is low; (5) land use compatible, provided special sound reinforcement systems are installed; (6) residential buildings require an NLR of 25 dBA; (7) residential buildings require an NLR of 30 dBA; (8) residential buildings not permitted.

Source: FAA (1989).

Federal Aviation Administration

Table 5-15 shows FAA land use compatibility guidelines, from Appendix A of FAR Part 150, *Airport Noise Compatibility Planning*. The final responsibility for interpreting this information rests with local authorities. Noise abatement measures offered include land acquisition, barrier construction, building soundproofing, implementation of a preferential runway system, use of flight procedures, and airport use restrictions (e.g., denial of use to aircraft not meeting noise standards, capacity limitations, use of approved noise abatement takeoff and landing procedures, landing fees, or curfews).

Federal Transit Administration

The FTA guidelines state that all reasonable steps must be taken to minimize the adverse effects of noise impacts. The first mitigation measure recommended is to explore alternate site locations that would eliminate the impact. If no feasible alternate sites exist, mitigation measures to explore would include enforcing equipment noise emissions specifications, treating rail wheels (e.g., using damping, resilient wheels, antilocking brakes, or maintenance), treating vehicles (e.g., using mufflers, shielding, or quiet fans), imposing operational restrictions, smoothing road and rail, erecting barriers, creating buffer zones by land acquisition, and building exterior soundproofing design.

Occupational Safety and Health Administration

In accordance with the Hearing Conservation Amendment, when 8-h TWA levels of 85 dBA (the action level) are exceeded, the OSHA requires that a hearing conservation program be implemented at the occupational facility. The principal ingredients of a hearing conservation program are outlined in Table 5-16.

TABLE 5-16   Main Ingredients of Hearing Conservation Programs

Program Component	Implementation Measures
Noise hazard identification	Sound level surveys including all levels between 80 and 130 dBA
Noise hazard evaluation	TWA of 85 dBA indicates necessity for hearing conservation program
	Post warning signs
	Suggest noise control measures
	Resurvey when levels change
Engineering, administrative controls	Prime targets are equipment emitting levels of 95 dBA or more
Hearing protection	Provide HPDs for employees capable of reducing employee TWA levels to 85 dBA for employees with STS; 90 dBA otherwise
Education and training	To include:
	Contents of pertinent noise standards
	Effects of noise on hearing
	Significance of controls for minimizing exposure to hazardous levels
	Purpose, selection, fitting, use, and care of HPDs
	Explanation of audiometric testing
Audiometric testing	Provide baseline audiogram within first 6 months of exposure above action level
	Notify employee with results
	Repeat annually until no exposure above action level

Source: 29 CFR 1910.95.

The main reasons for a hearing conservation program are to protect workers from noise-induced hearing loss, to comply with OSHA regulations (and thus avoid fines that are levied if no action has been taken to reduce noise levels into compliance with the regulations or to set up a hearing conservation program within 60 days of an OSHA inspection finding the facility in violation of the regulations), and to minimize the obligation of the employer to pay hearing loss compensation.

A facility must first be surveyed for high-noise areas. This is typically accomplished by plotting constant noise-level contours, as are shown in the examples in Figs. 5-5 through 5-7, and/or examining dosimetry results. As a minimum requirement, all noise between 80 and 130 dBA must be included in the noise survey. When noise levels exceeding 85 dBA are revealed, it is preferable that engineering controls be designed to reduce levels to below 85-dBA wherever employees would be working. The most common way of dealing with the problem, however, is the least expensive way of posting warning signs at the boundary between SPLs above and below 85 dBA (on the 85 dBA contour line from the noise survey). These warning signs would state that HPDs are required to be worn in the designated areas. Because the OSHA is concerned only with the noise level that reaches the inner ears of employees, the noise attenuation provided by an HPD is often enough to reduce unacceptable noise levels outside the ear to acceptable levels inside.

Audiometric (hearing threshold) testing is a necessary part of the hearing conservation program to ensure that no noise-induced hearing loss results from an employee being present in the work environment. Baseline audiograms (hearing sensitivity tests) are required to be performed before an employee begins working with a company, to use as a comparison with later audiograms. In this way, if an employee has a hearing loss identified before commencing his work with a company, it can be proven that this loss was not caused by employment with that company. Recreational activities outside the workplace (e.g., hunting or target shooting with firearms, or listening to excessive amounts of amplified music) can also cause noise-induced hearing loss. In addition, the drugs and diseases mentioned in Chapter 1 can cause hearing loss; many factors must therefore be ruled out before the diagnosis of occupational noise-induced hearing loss can be concluded.

The standard threshold shift (STS) is also known as the temporary threshold shift (TTS). The STS is defined as an average loss in hearing sensitivity of 10 dB at 2000, 3000, and 4000 Hz in either ear occurring directly following noise exposure. Hearing sensitivities usually return to normal after several hours. Baseline audiograms are to be given at least 14 h after an employee has been exposed to workplace noise to account for this. Presbycusis effects are also supposed to be taken into account in audiometric testing.

For comparison it can be noted that European countries, such as the United Kingdom (UK), use a 3-dBA exchange rate (the amount that the TWA must increase to cut the maximum allowable exposure time in half) whereas the OSHA uses a 5-dBA exchange rate. Also of note is that the maximum allowable exposure level for any period of time is 105 dBA for the UK and 115 dBA for the OSHA.

American Public Transit Association

The criteria from *Guidelines for Design of Rapid Transit Facilities* (APTA, 1981) provide limits that are expected to be achievable in practice for systems under proper maintenance and design. Therefore, minimal mitigation information is provided outside of some noise control recommendations to keep the rail noise within the limits. To minimize rail–wheel squeal, the wheels can be treated with damping materials or replaced with resilient wheels.

Department of Defense

Applicable local and federal agencies are apprised of the AICUZ study information (USAF, 1992). Solutions to potential noise problems are worked out between the DoD and these agencies. Mitigation measures are limited only by the restrictions of the individual communities and safety considerations for aircraft operations.

STATE NOISE REGULATIONS AND GUIDELINES

Agencies

The only regulations and guidelines applicable to environmental noise assessment in every state are those promulgated by the respective state departments of transportation for vehicular traffic noise. Regulations and guidelines for rail, aircraft, and stationary sources are usually handled on a federal or local level. Some states have their own noise codes that are usually in the same format as local noise ordinances discussed in Sample Municipal Noise Codes and Ordinances (below).

Noise Assessment Methods

Each state's department of transportation (DOT), as directed by the FHWA, uses the FHWA vehicular traffic regulations with its own interpretations of impact definition.

Descriptors

Most states use the $L_{eq(1)}$ as the choice of descriptor under FHWA regulations for vehicular traffic noise. There are some that still use the $L_{10(1)}$ descriptor.

Some states also have a department of environmental protection (DEP) that has established noise regulations. For example, the New Jersey DEP established the New Jersey Noise Control Code as Title 7, Chapter 29, Subchapter 1 of the New Jersey Administrative Code in 1984 (revised in 1985). The descriptors chosen for this code are maximum instantaneous SPL, using the slow meter response, in terms of overall dBA or octave band levels.

The state of California, in addition to the DOT regulations in compliance with the FHWA, has four codes that cover different noise sources and scenarios. The California State Noise Insulation Standards are included in the State Building Code (Part 2, Title 24, California Code of Regulations, published December 1988). This is discussed in Chapter 7. The California Motor Vehicle Laws, from the California Vehicle Code, have sections adopted over different years through the 1970s. Their format is similar to that used by the EPA for their noise regulations with instantaneous SPL limits. The aircraft noise standards, in terms of CNEL, are included in Title 21, Subchapter 6, California Code of Regulations (CCR), published March 1990. The *State of California General Plan Guidelines* (Cervantes et al., 1990) offers guidelines to assist local municipalities in drafting noise codes in accordance with Government Code Section 65302(f) to address a noise element as part of the general plan of a municipality. A choice of L_{dn} or CNEL is offered.

Study Area

All state DOT regulations are based on FHWA regulations and study area considerations are thus the same as those promulgated by the FHWA. The New Jersey DEP defines the study area beginning at the closest residential or commercial property line to the source in question.

Models and Analysis Techniques

All state transportation agencies are required to use FHWA regulations with their own choice of descriptor and interpretations of impact criteria. Therefore, the noise prediction methodologies used by states are based on the FHWA models.

The Pennsylvania Department of Transportation (PennDOT) has detailed documentation on noise impact assessment procedures (PennDOT, 1987) that is worth reviewing for reference. Three levels of analysis can be explored:

qualitative, screening, or detailed. A qualitative analysis, consisting of a narrative discussion of the project and its relationship to noise-sensitive locations, is performed on projects determined to have noise impacts that are not significant but cannot be exempted from analysis. A screening analysis, to be performed for projects that are not complex in nature and have few sensitive locations, would involve using the FHWA hand calculation method. However, if levels predicted from this analysis approach or exceed the FHWA criteria, a detailed analysis is required.

The detailed analysis requires using STAMINA 2.0 methodology for predicting future noise levels and for comparison with measured existing noise levels. Noise measurements are to be performed during the worst-case hour. When a condition is proposed in which noise level prediction is not appropriate or possible (such as for tunnels, covered roadways, or parking lots), another location with similar characteristics can be monitored for data. Similar procedures are outlined for construction noise assessment.

Acceptable Noise Levels According to State Agencies

New Jersey Department of Environmental Protection
The New Jersey Department of Environmental Protection (NJDEP) provides not-to-exceed instantaneous noise level limits for different categories in its noise code. The basis for these limits is not stated in the Administrative Code.

California
The California aircraft noise standards (Title 21, Subchapter 6, CCR, 1990) identify a CNEL of 65 dBA as the maximum acceptable level. The CNEL 65-dBA contour line is known as the noise impact boundary. An interior limit of 45 dBA CNEL is set for noise-sensitive locations.

The *State of California General Plan Guidelines* (Cervantes et al., 1990) provides land use compatibility criteria in terms of L_{dn} or CNEL. These criteria are similar in format to those established by HUD; however, the noise levels are different and overlap in many cases. The overlapping means that the acceptability of the land use would depend on its specific use and sensitivity. The four noise level categories are normally acceptable, conditionally acceptable, normally unacceptable, and clearly unacceptable. Conditionally acceptable means that noise reduction must be considered but conventional construction with closed windows and year-round air conditioning should suffice. Normally unacceptable means that development should proceed only if proper noise insulation designs are included. Table 5-17 shows the specific recommendations.

TABLE 5-17   California Land Use Compatibility Guidelines[a]

Land Use Category	Normally Acceptable	Conditionally Acceptable	Normally Unacceptable	Clearly Unacceptable
Residential, low density	<60	55–70	70–75	>75
Residential, multifamily	<65	60–70	70–75	>75
Transient lodging	<65	60–70	70–80	>80
Schools, libraries, churches, hospitals, nursing homes	<70	60–70	70–80	>80
Auditoriums, concert halls, amphitheaters	No rating	<70	No rating	>65
Sports arena, outdoor spectator sports	No rating	<70	No rating	>70
Playgrounds, parks	<70	No rating	67–75	>72
Golf courses, riding stables, water recreation facilities, cemeteries	<75	No rating	70–80	>80
Office buildings, commercial and professional	<70	67–77	75–85	No rating
Industrial, agricultural, manufacturing, utilities	<75	70–80	75–85	No rating

[a] In dBA L_{dn} or CNEL.

Source: Cervantes et al. (1990).

Impact Criteria According to State Regulations and Guidelines

Impact criteria for state DOT agencies are summarized in Table 5-18 for those that responded to a recent FHWA survey (FHWA, 1991).

As Pennsylvania has a unique graduated scale, it is described briefly.

Pennsylvannia Department of Transportation

The Pennsylvania Department of Transportation (PennDOT) uses FHWA regulations and chooses the $L_{eq(1)}$ noise descriptor for impact definition, using the NAC criteria. The relative impact of a substantial increase over existing levels is combined with the absolute impact criteria and divided into three scenarios for FHWA land use category B (including residential and most common noise-sensitive locations).

1. When the total predicted future $L_{eq(1)}$ noise levels with the proposed project equal or exceed 66 dBA, abatement considerations are required regardless of the increase or decrease in noise levels compared to existing noise levels.
2. When the total predicted future $L_{eq(1)}$ noise levels with the proposed project are between 57 and 66 dBA, abatement considerations are required

TABLE 5-18   State Department of Transportation Interpretations of Federal Highway Administration Noise Regulations

State	Descriptor	"Approach or Exceed" Interpretation	"Substantial Increase" Interpretation
California	L_{eq}	Equals NAC	12-dBA increase and ≥ 65 dBA
Connecticut	L_{10}	Equals NAC	15 dBA or $>$ NAC
Florida	L_{eq}	Within 2 dBA of NAC	≥ 10–15 dBA
Illinois	L_{eq}	Equals NAC	14-dBA increase or within 2 dBA of NAC and $>$ 12-dBA increase
Kentucky	L_{eq}	Equals NAC	≥ 10 dBA
Maryland	L_{eq}	Equals NAC	≥ 10 dBA
Massachusetts	L_{eq}	66 dBA	> 15 dBA
Michigan	L_{eq}	Equals NAC	> 10 dBA
Minnesota	L_{10}	Equals NAC	≥ 10 dBA
Missouri	L_{eq}	Equals NAC	> 10 dBA
Montana	$L1_{eq}$	Equals NAC	≥ 10 dBA
New Jersey	L_{eq}	64 dBA	≥ 10 dBA or > 64 dBA
New Mexico	L_{eq}	Equals NAC	≥ 10 dBA and > 57 dBA
New York	L_{eq}	Equals NAC	≥ 6 dBA
North Carolina	L_{eq}	Within 2 dBA of NAC	≥ 10 dBA
North Dakota	L_{eq}	Within 3 dBA of NAC	No response
Oklahoma	L_{10}	Equals NAC	No response
Oregon	L_{eq}	Equals NAC	≥ 10 dBA
Pennsylvania	L_{eq}	66 dBA	≥ 10–15 dBA
Puerto Rico	L_{eq}	66 dBA	≥ 10 dBA
Rhode Island	L_{eq}	Equals NAC	No response
Texas	L_{eq}	Equals NAC	≥ 10 dBA and > 65 dBA
Utah	L_{eq}	Within 2 dBA of NAC	≥ 10 dBA
Vermont	L_{eq}	Equals NAC	≥ 10 dBA
Virginia	L_{eq}	Equals NAC	≥ 10 dBA

Source: FHWA (1991).

when increases of 10 to 15 dBA over existing levels would occur. Substantial increases of 10 dBA apply to a resulting total $L_{eq(1)}$ level of 66 dBA whereas a 15-dBA increase applies to a predicted total $L_{eq(1)}$ level of 57 dBA. Values corresponding to substantial increases between 57 and 66 dBA are interpolated between 10 and 15 dBA.

3. When the total predicted future $L_{eq(1)}$ noise levels with the proposed project are equal to or less than 57 dBA, abatement considerations are not warranted, regardless of the increase.

For FHWA land use category A (the most sensitive uses to noise), noise abatement consideration is required whenever the total predicted future $L_{eq(1)}$ levels with the proposed project approach or exceed 57 dBA. The other FHWA land use categories (C, D, and E) are reviewed on an individual basis with no set criteria.

New Jersey Department of Environmental Protection
The New Jersey Department of Environmental Protection (NJDEP) sets forth the following noise limits in its *Noise Code* (Title 7, Chapter 29, Subchapter 1, NJ Administrative Code, 1985):

For noise generated by industrial, commercial, public service, or community service facilities at another commercial facility at any time or at a residential property line between 7:00 A.M. and 10:00 P.M., instantaneous noise levels shall not exceed:

1. 65 dBA, or
2. any of the octave band levels listed below:

Octave Band Center Frequency (Hz)	SPL (dB)
31.5	96
63	82
125	74
250	67
500	63
1000	60
2000	57
4000	55
8000	53

3. an impulsive level of 80 dB

For noise generated by industrial, commercial, public service, or community service facilities at a residential property line, instantaneous noise levels, from 10:00 P.M. to 7:00 A.M. shall not exceed:

1. 50 dBA, or
2. any of the octave band levels listed below:

Octave Band Center Frequency (Hz)	SPL (dB)
31.5	86
63	71
125	61
250	53
500	48
1000	45
2000	42
4000	40
8000	38

3. an impulsive level of 80 dB

California

The California Vehicle Code noise limits are given in Table 5-19.

The noise impact area, according to the California aircraft noise standards, is the area inside the 65-dBA CNEL contour that contains noise-sensitive locations. The *State of California General Plan Guidelines* (Cervantes et al., 1990) do not discuss specific impact criteria but leave that to the individual municipalities to decide, based on their specific environment and interests.

Mitigation Requirements

State Department of Transportation Agencies

All state DOT agencies are required to follow the FHWA regulations for mitigation of noise impacts. PennDOT, in addition to utilizing the FHWA traffic mitigation options mentioned above, lists mitigation options for construction noise impacts. These include design considerations (e.g., locating the project site where artificial and natural barriers and buildings provide shielding to noise-sensitive areas), sequence of operations (e.g., performing several steps concurrently or erecting planned highway noise barriers before other construction to provide construction noise shielding during construction), source control (i.e., specifying quiet equipment), site control (e.g., locating stationary equipment as far as possible from sensitive areas, enclosing areas, or erecting temporary barriers), and time and activity constraints (e.g., limiting activities to daytime hours).

TABLE 5-19   California Vehicle Code Noise Limits[a]

Noise Source	Conditions	SPL Limit
Motor vehicle		
> 10,000 lb	Any conditions, < 35 mph	86
	Any conditions, > 35 mph	90
> 6000 lb	Level grade, < 35 mph constant	82
Motorcycle	Any conditions, < 45 mph	82
	Any conditions, > 45 mph	86
	Level grade, < 35 mph constant	77
Any other motor vehicle	Any conditions, < 45 mph	76
	Any conditions, > 45 mph	82
	Level grade, < 35 mph constant	72
Off-highway vehicle	EPA-type test, after model 1974	86

[a] In dBA at 50 ft.

Source: California Vehicle Code as of 1992.

New Jersey Department of Environmental Protection

The NJDEP provides not-to-exceed noise limits but does not discuss mitigation methods. If the published limits are exceeded, fines can be issued until the noise levels are within the limits.

California

The California aircraft noise standards suggest several methods for controlling and reducing airport noise. These include encouraging quieter and discouraging noisier aircraft from using the facility, encouraging flight paths and procedures that minimize noise over sensitive areas, establishing runway utilization planning schedules that are most compatible with the area, reducing flight operations, using barriers, rezoning, acquiring incompatible land, and applying acoustic insulation to affected noise-sensitive uses.

The *State of California General Plan Guidelines* (Cervantes, 1990) recommend general mitigation measures that need to be specialized to the individual communities. These measures include sound insulation, enforcing noise emissions standards, minimizing siren usage, an animal philosophy (e.g., dealing with barking dogs) of treating the problem rather than issuing citations, and adopting ordinances that deal with the specific problems of each community.

SAMPLE MUNICIPAL NOISE CODES AND ORDINANCES

Cities and townships deal with noise issues through noise codes, regulations, and ordinances. Samples of city noise regulations are described below. These include the noise regulations of New York City, Los Angeles, and Chicago. Because the criteria of New York City involve the NEPA type of environmental assessment process, their criteria will be discussed in the same format as was used above for the state and federal agencies. The Los Angeles and Chicago codes are closer to the formats of local ordinances and are discussed in a more general format.

New York City Noise Regulations

New York City has three regulations, each promulgated by a different authority and dealing with a different type of noise source. These three regulations are the New York City Noise Code, the New York City Zoning Resolution, and the New York City CEPO-CEQR Standards.

New York City Noise Code

The New York City Noise Code was adopted in 1972 as Title 24, Chapter 2 of the New York City Administrative Code. Noise limits for motor vehicles, air compressors, circulation devices, claxons (horns), emergency signal devices, paving breakers, and commercial music are listed in the code. Unnecessary noise is prohibited and three ambient noise quality zones are established by zoning designations and rated in terms of noise limitations. Construction activities are also limited in the code. Noise from highways, rail, and aircraft operations are not included in the code.

New York City Zoning Resolution
The New York City Zoning Resolution, in 1960, adopted Section 42-21 as noise performance standards for any activity in manufacturing districts. Different limits are presented for each type of manufacturing district and correction factors are offered for adjacent residential districts and impulsive noise.

CEPO-CEQR Standards
The New York City Department of Environmental Protection (NYCDEP), Office of Environmental Impact (OEI), in 1983, adopted City Environmental Protection Order-City Environmental Quality Review (CEPO-CEQR) noise standards for environmental impact review. Four categories of acceptability are established, based on noise level limits and land use, for vehicular traffic, rail, and aircraft noise sources. Only mobile sources are included in the standards. Each of the three noise source classifications is analyzed separately and in terms of different descriptors. Mitigation requirements are listed according to the noise category. Both absolute and relative impact criteria are presented.

Descriptors
The New York City Noise Code uses $L_{eq(1)}$ as the descriptor for its ambient noise quality zone criteria. Other descriptors used are maximum instantaneous SPL (in dBA) for motor vehicles, air compressors, claxons, emergency signal devices, and paving breakers outdoors and circulation devices and commercial music indoors, and maximum instantaneous 1/3-octave band SPL (in dB) for commercial music indoors.

The New York City Zoning Resolution uses maximum instantaneous octave band SPL (in dBC), using the slow meter response, as its noise descriptor for industrial noise sources.

The CEPO-CEQR Standards use $L_{10(1)}$ for vehicular traffic sources, daily L_{dn} for rail sources, and the annual average L_{dn} for aircraft sources. Each mobile source is analyzed separately, using its own assigned descriptor.

Typical municipal noise ordinances use maximum instantaneous overall SPL (in dBA) as the noise descriptor for environmental assessment of sources other than public transportation. Public transportation sources are usually regulated on the federal, state, or specific agency level.

Study Area

The New York City Noise Code specifies minimum distances from noise sources, different for each type of source, as the beginning of study areas. For ambient noise quality zone analysis, the closest zone boundary to the source in question is the beginning of the study area. For specific sources, a distance of 1 m is specified for air compressors and paving breakers, 10 ft from the center line is specified for refuse compacting vehicles, 25 and 50 ft from the center line of the rear face of motor vehicles (for non-public highway travel) or the center of the lane of the public highway, 50 ft from the center of the front face of vehicles having claxons or emergency signal devices, or 3 ft from an open window inside a residence for commercial music or circulation devices. Unnecessary noise is also defined at any noise-sensitive location.

The New York City Zoning Resolution defines the study area beginning at the lot line of the manufacturing district in question.

The study area for CEPO-CEQR Standards is typically interpreted as any area having noise-sensitive locations that could be affected by project-generated vehicular traffic, rail, aircraft, or construction noise.

Typical municipal ordinances set not-to-exceed instantaneous limits at specific land use property lines.

Noise Analysis Methods

Neither the New York City Noise Code nor the New York City Zoning Resolution refers to assessment methodologies. They only provide absolute limits for noise levels. The only guidance for noise measurements provided is deferral to ANSI standards in the New York City Noise Code and specification of using C-weighting and slow response speed for an octave band analyzer in the New York City Zoning Resolution. Also specified is the use of an impact noise analyzer for monitoring impulsive signals.

CEPO-CEQR Standards do not specify methodologies or analysis techniques. A proportional prediction methodology is usually used for vehicular traffic noise analysis. This technique compares predicted future No Build and Build traffic volumes with existing traffic volumes and extrapolates noise level increases as the logarithm of the ratio of traffic volumes. Traffic volumes are usually expressed in terms of passenger car equivalent (PCE) values, as defined by FHWA and discussed at the beginning of this chapter. Prediction schemes for rail and aircraft sources are not offered. Measurement techniques are deferred to ANSI standards. Measurements are to be performed during the worst-case hour (when project-generated noise increases would be the largest) and 20 min is usually considered acceptable for extrapolation to hourly readings when no loud uncharacteristic events are occurring. Traffic counts (having vehicle classifications recorded concurrently with noise monitoring) are suggested but not required.

Acceptable Levels

The New York City Noise Code provides not-to-exceed instantaneous and $L_{eq(1)}$ limits for different categories of outdoor and indoor environments. It is implied in the code that interior instantaneous noise levels less than 45 dBA would provide an acceptable interior environment. Exterior limits vary with circumstances and sources.

The New York City Zoning Resolution provides not-to-exceed instantaneous noise limits for the three manufacturing district categories. No basis is stated for these limits in the New York City Zoning Resolution.

The NYCDEP provides four levels of acceptability for six categories of land use for vehicular traffic, rail, and aircraft sources. The acceptability categories are generally acceptable, marginally acceptable, marginally unacceptable, and clearly unacceptable. These categories and associated noise limits are for exterior levels only. Generally acceptable levels are shown in Table 5-20. The exterior limitations are based on an interior maximum noise level ($L_{10(1)}$ or L_{dn}, depending on the source) goal of 45 dBA, assuming that exterior window/wall noise attenuation, using typical construction materials, is 20 to 25 dBA.

TABLE 5-20   New York City CEPO-CEQR Exterior Noise Exposure Standards[a]

Noise Receptor Classification	Time	Acceptable	Marginally Acceptable	Marginally Unacceptable	Clearly Unacceptable
Outdoor areas, quiet	All hours	$L_{10} \leq 55$			
Hospitals, nursing homes	All hours	$L_{10} \leq 55$	$55 < L_{10} \leq 65$	$65 < L_{10} \leq 80$	$L_{10} > 80$
Residential	7:00 A.M.–11:00 P.M.	$L_{10} \leq 65$	$65 < L_{10} \leq 70$	$70 < L_{10} \leq 80$	$L_{10} > 80$
	11:00 P.M.–7:00 A.M.	$L_{10} \leq 55$	$55 < L_{10} \leq 70$	$70 < L_{10} \leq 80$	$L_{10} < 80$
Other public gathering places, commercial offices	All times	$L_{10} \leq 65$	$65 < L_{10} \leq 70$	$70 < L_{10} \leq 80$	$L_{10} < 80$
Airport, train environments	All times	$L_{dn} \leq 60$	$60 < L_{dn} \leq 65$	$65 < L_{dn} \leq 75$	$L_{dn} > 75$

[a] In dBA.

Source: NYCDEP (adopted policy, 1983).

Impact Criteria

The New York City Noise Code sets forth instantaneous noise level limits on specific sources and worst-hour L_{eq} limits for ambient noise quality zone (ANQZ) criteria. All levels set forth are not-to-exceed limits.

In terms of specific sources, the code limits noise levels according to Table 5-21.

In terms of ANQZ, Table 5-22 lists the limits that apply to noise generated by stationary sources within the boundaries of a project. Noise from construction sources is not included in the ANQZ criteria.

The New York City Zoning Resolution sets forth instantaneous dBC noise level limits for manufacturing use property lines. These limits are in terms of so-called octave band frequencies, yet the frequency designations do not agree with standard octave band representation. Table 5-23 shows the limits as stated in Section 42-213 of the regulations.

For impulsive sounds, the limits in Table 5-23 are increased by 6 dB.

TABLE 5-21   New York City Noise Code Instantaneous Noise Limits[a]

Noise Source	Distance[b]	SPL Limit
Motor vehicle		
>8000 lb GW, <35 mph	50	86
>8000 lb GW, >35 mph	50	90
>8000 lb GW, <35 mph	25	92
>8000 lb GW, >35 mph	25	96
Motorcycle		
<35 mph	50	78
>35 mph	50	82
<35 mph	25	84
>35 mph	25	88
Any other motor vehicle		
<35 mph	50	70
>35 mph	50	79
<35 mph	25	76
>35 mph	25	85
Air compressor	1 m	80
Circulation device indoors	3 ft from open window	45
Refuse-compacting vehicle	10	70
Motor vehicle claxon	50	98
Emergency signal device	50	90
Paving breaker	1 m	95
Commercial music indoors (source outside the space in question)	Any	45 dBA or 45 dB in any 1/3-octave band, 63–500 Hz

[a] In dBA.

[b] In feet, unless otherwise indicated.

Source: New York City Noise Code (1972).

TABLE 5-22   City of New York Ambient Noise Quality Zone Criteria[a]

Ambient Noise Quality Zone (ANQZ)	Daytime Standards (7:00 A.M.–10:00 P.M.)	Nighttime Standards (10:00 P.M.–7:00 A.M.)
N1: Low-density residential (R1 to R3) land use	60	50
N2: High-density residential (R4 to R10) land use	65	55
N3: Commercial and manufacturing (C1 to C8, M1 to M3) land use	70	70

[a] $L_{eq(1)}$ in dBA.

Source: New York City Noise Code (1972).

Whenever a manufacturing district is adjacent to a residential district, the limits in Table 5-23 are reduced by 6 dB.

The NYCDEP considers a project-generated increase (between the No Build and Build conditions) of 3 dBA or more at a noise-sensitive location to be a significant adverse noise impact. This noise level increase would correspond to the descriptor being used (i.e., $L_{10(1)}$ for vehicular traffic, daily L_{dn} for rail, and annual average L_{dn} for aircraft). Mitigation of impacts depends on the specified noise category of the absolute levels after the proposed project has been built and is operational.

Mitigation Requirements

Noise impact mitigation requirements as promulgated by the New York City Noise Code, the New York City Zoning Resolution, and CEPO-CEQR Standards are discussed below.

TABLE 5-23   New York City Zoning Resolution Maximum Permitted SPL[a]

Octave Band (Hz)	District		
	M1	M2	M3
20–75	79	79	80
75–150	74	75	75
150–300	66	68	70
300–600	59	62	64
600–1200	53	56	58
1200–2400	47	51	53
2400–4800	41	47	49
Above 4800	39	44	46

[a] In dBC.

Source: New York City Zoning Resolution (1960).

The New York City Noise Code provides not-to-exceed noise limits and prohibits unnecessary noise. Construction activities are limited to weekdays between 7:00 A.M. and 6:00 P.M. Mitigation measures are discussed only in the terms that mufflers and the latest noise control designs should be incorporated into activities and devices. If the published limits are exceeded, fines can be issued until the noise levels comply with the code.

The New York City Zoning Resolution provides not-to-exceed noise limits but does not discuss mitigation methods or requirements.

Table 5-24 shows the mitigation requirements of CEPO-CEQR standards in terms of the exterior level noise category and source (vehicular traffic, rail, or aircraft). Mitigation is required whenever a significant impact (i.e., a project-generated $L_{10(1)}$ or L_{dn} increase of 3 dBA or more) is expected.

The attenuation values required in Table 5-24 are for the exterior walls of residential buildings. Commercial office spaces would require 5 dBA less attenuation than that listed in Table 5-24.

In addition, the NYCDEP has published construction noise mitigation guidelines. This document, entitled *Construction Noise Mitigation Measures*, was published as DNA (Division of Noise Assessment) Report #CON-79-001 in July 1979. Mitigation measures are divided into four categories: source control, site control, time and activity constraints, and traffic considerations. Source control includes using the quietest equipment and methods (e.g., using a single piece of equipment instead of many small ones, using wheeled instead of tracked vehicles, using welding instead of riveting, mixing concrete off site instead of on site, and using prefabricated structures instead of assembling them on site), using electrical equipment instead of those powered by internal combustion engines, using mufflers and shields, using vibratory instead of conventional pile drivers, treating pile drivers with resilient pads or dampening materials, erecting barriers and enclosures, and proper equipment maintenance. Site control includes locating equipment as far as possible from noise-sensitive areas and using barriers. Time and activity constraints include scheduling noisy operations simultaneously, limiting noisy activities near schools to before 9:00 A.M. and after 3:00 P.M. or when school is not in session, and using the New York City Noise Code restrictions of working on weekdays only between 7:00 A.M. and 6:00 P.M.. Traffic considerations include routing of construction and affected road traffic to minimize noise impacts.

The NYCDEP also has provisions that developers will attempt to assure that the half-hour L_{eq} would be ≤ 75 dBA at the nearest residential property line and ≤ 80 dBA at the closest commercial building during construction activity. Property line SPL measurements are required to be performed on a monthly basis and the results are to be compared with estimated off-site sound levels to assess the effectiveness of the noise control measures.

As possible mitigation measures, project alternatives can be developed.

TABLE 5-24 New York City CEPO-CEQR Noise Attenuation Guidelines[a]

Noise Category	Marginally Acceptable	Marginally Unacceptable		Clearly Unacceptable		
		I	II	I	II	III
Attenuation required	25	30	35	40	45	50
Vehicular noise	$65 < L_{10} \leq 70$	$70 < L_{10} \leq 75$	$75 < L_{10} \leq 80$	$80 < L_{10} \leq 85$	$85 < L_{10} \leq 90$	$90 < L_{10} \leq 95$
Train noise	$60 < L_{dn} \leq 65$	$65 < L_{dn} \leq 70$	$70 < L_{dn} \leq 75$	$75 < L_{dn} \leq 80$	$80 < L_{dn} \leq 85$	$85 < L_{dn} \leq 90$
Aircraft noise	$60 < L_{dn} \leq 65$	$65 < L_{dn} \leq 70$	$70 < L_{dn} \leq 75$	$L_{dn} > 75$	NA	NA

[a] In dBA.

Source: NYCDEP (adopted policy, 1983).

In developing project alternatives to reduce or avoid noise impacts, the simplest and most common way of analyzing the situation is to calculate the conditions that would just avoid an impact and tailor the alternative project to that new scenario. For instance, if a CEPO-CEQR vehicular traffic noise impact were identified at a noise-sensitive location, the project-generated L_{10} worst-hour increase would be at least 3 dBA. If one calculated the project-generated traffic volume that would cause a 2.9-dBA increase in worst-hour L_{10} values, that traffic volume would define the alternative project volume. To accomplish this, vehicles may have to be rerouted onto other thoroughfares or the project may have to be scaled down. If traffic is rerouted, analysis must be performed to ensure that this would not create new impacts at different noise-sensitive locations. Similar analysis techniques to this can be used for analyzing alternatives from any relative impact criterion.

When dealing with absolute impact criteria, alternative project arrangements can be set by moving, scaling down, or shielding the original project to the point where impacts are avoided. For instance, if a manufacturing facility generated a significant impact at a residence by exceeding the New York City Zoning Resolution, the noise-generating part of the facility could be moved to the distance at which the noise levels at the property line would conform with the New York City Zoning Resolution. Another possible alternative would be to scale down operations such that noise levels would not exceed the prescribed limits at the property line. Yet another alternative to the project could include a building between the noise-generating facility and the property line to shield the noise to the point at which an impact would be avoided. These options would each have to be evaluated in terms of their feasibility and potential impacts on new areas.

Los Angeles Noise Regulations

The Los Angeles Noise Regulation (Chapter XI of the City Code, Revision Number 33, published in 1987) is based on whether monitored noise levels, averaged over at least 15 min without including occasional uncharacteristic loud events, exceed the levels listed in Table 5-25 at specified locations. This format is similar to that used for the New York City Ambient Noise Quality Zone criteria. Most noise sources are restricted by hours of operation and whether or not they are audible at residential locations. The specific types of noise sources regulated are sound reproductive devices, HVAC (heating, ventilating, and air conditioning) equipment, construction sources, powered equipment, sanitary operations, vehicular sources, and general noise.

At the boundary between two zones, the lower limit is assumed. In addition, any noise levels measured from a steady tone above 200 Hz in

TABLE 5-25   Los Angeles Average Noise Level Limits[a,b]

Zone	Daytime (7:00 A.M.–10:00 P.M.)	Nighttime (10:00 P.M.–7:00 A.M.)
A1, A2, RA, RE, RS, RD, RW1, RW2, R1, R2, R3, R4, R5 (residential)	55	45
P, PB, CR, C1, C1.5, C2, C4, C5, CM (commercial)	65	60
M1, MR1, MR2 (manufacturing)	70	70
M2, M3 (industrial)	75	75

[a] In dBA.

[b] These limits are based on prescribed minimum background levels that are 5 dBA less than all values listed in the table. If background levels are higher than these prescribed values, the limits to be used are 5 dBA higher than the measured background levels.

Source: Los Angeles Noise Code (1987).

frequency or a repeating impulsive source should have 5 dBA added to the noise reading before comparing with the limits in Table 5-25; and any noise occurring for less than 15 min in any consecutive 60-min period between 7:00 A.M. and 10:00 P.M. should have 5 dBA subtracted from the noise reading before comparing with the limits of Table 5-25.

Specific Limitations
Following are the specific restrictions of the Los Angeles Noise Code. These limitations do not apply in emergency situations.

Sound Reproductive Devices
Audible noise levels are prohibited from sound reproductive devices either 150 ft from the property line of the source within a residential zone or within 500 ft of a residential zone. Also, noise levels caused by such devices in adjoining units, residential or business, that exceed the values in Table 5-25 are prohibited. In addition, the operation of sound amplification systems is prohibited between 10:00 P.M. and 7:00 A.M. For noncommercial purposes in residential zones or within 500 ft of them, this equipment is prohibited from use between 4:30 P.M. and 9:00 A.M. For commercial purposes in nonresidential zones or within 500 ft of them, this time of prohibition is from 9:00 P.M. to 8:00 A.M. The only sounds permitted are human speech and music at any time.

Heating, Ventilating, and Air-Conditioning Equipment
Included in this category are pool and reservoir maintenance equipment. For this category, Table 5-25 limits are used for any occupied property.

Construction Sources

General construction activities that would cause disturbing noise to occupants are prohibited between 9:00 P.M. and 7:00 A.M. In addition, no person (other than an individual working on his or her own single family dwelling) can perform any construction activity within 500 ft of a residential property on Saturdays between 6:00 P.M. and 8:00 A.M. or any time on Sundays.

Powered Equipment

The operation of any powered equipment that generates loud noise is prohibited between 10:00 P.M. and 7:00 A.M. in any residential zone or within 500 ft of one. Also, when operating, the equipment cannot generate noise levels that exceed the limits of Table 5-25 in any occupied unit. In addition, maximum instantaneous noise levels of industrial or construction powered equipment cannot exceed 75 dBA at 50 ft and levels of residential lawn maintenance equipment cannot exceed 65 dBA at 50 ft.

Sanitary Operations

Garbage collection and disposal is prohibited within 200 ft of any residential building between 9:00 P.M. and 6:00 A.M.

Vehicular Sources

Vehicles cannot be repaired within 500 ft of a residential zone between 8:00 P.M. and 8:00 A.M. in such a manner that the noise generated is audible at least 150 ft from the property line of the activity or that the noise generated is within the Table 5-25 limits at any occupied residential property. These same limits apply to vehicles driving on a property at any time and sounding horns or accelerating unreasonably. Any signaling device that can be heard more than 200 ft away is also prohibited. The code does not apply to vehicles on public highways or streets.

Vehicle theft alarm systems are limited to be silenced within 5 min of the first emission of audible sound. Also, any vehicle loading activity that causes unnecessary noise is prohibited between 10:00 P.M. and 7:00 A.M. within 200 ft of residential buildings.

General Noise

A provision is made in the Los Angeles Noise Code that prohibits "loud, unnecessary, and unusual noise which disturbs the peace or quiet of any neighborhood or which causes discomfort or annoyance to any reasonable person of normal sensitiveness residing in the area." This would cover any noise source not mentioned in the code and requires a judgment call by the Los Angeles Police Department.

Chicago Noise Regulations

The Chicago Noise Regulation (Title 11, Article VII of the City Municipal Code, adopted January 27, 1988) is based on maximum instantaneous sound pressure level limits and time constraints. As for the Los Angeles Noise Code, this regulation does not apply to emergency situations, mass transit sources, and aircraft sources. Specific sources regulated are motor vehicles, powered equipment, and manufacturing districts. Similar to the New York City Zoning Resolution noise limits, the Chicago Noise Code limits noise from manufacturing districts according to maximum SPLs at the district boundaries in terms of octave band levels.

The general exterior noise limitations are that, in any environment, no person can generate noise levels either in excess of 80 dBA at 10 ft or such that the noise is audible at least 600 ft from the source. The interior limitations are that no person can generate noise levels exceeding 55 dBA within any residential unit between 9:00 P.M. and 8:00 A.M. Beside the typical exemptions from the Code, construction noise is exempt between 8:00 A.M. and 9:00 P.M.

Generally prohibited acts include sounding amplified signals that create a disturbance for more than 5 min in 1 h, intentionally sounding alarms for nonemergency or nontesting purposes, generating noise that interferes with the functions of any school, library, nursing home, or medical facility, causing a noise disturbance at a noise-sensitive location from loading or sanitary operations between 10:00 P.M. and 7:00 A.M, blowing steam whistles as a signal for commencing or suspending work or for any other nonemergency purposes, and using construction equipment between 9:00 P.M. and 8:00 A.M. within 600 ft of any residential building or hospital.

Specific Limitations

Motor Vehicles
It is prohibited for any motor vehicle with a gross weight rating of more than 10,000 lb to operate while standing on a private property within 150 ft of a residential property for more than two consecutive minutes unless it is in a complete enclosure. In terms of new vehicle sales, for 50-ft noise limits under testing procedures similar to those used for EPA regulations, motorcycles cannot exceed 78 dBA, passenger cars cannot exceed 75 dBA, and vehicles (including buses) having gross weight ratings greater than 10,000 lb cannot exceed 77 dBA. In addition, under the same conditions as stated above, any motor vehicle or combination of vehicles towed by such vehicles cannot exceed 70 dBA when traveling 35 mph or less and 79 dBA when traveling over 35 mph. Any motorcycle cannot exceed 78 dBA when traveling 35 mph or less and 82 dBA when traveling over 35 mph.

New recreational and off-highway vehicles, such as snowmobiles, dune buggies, all-terrain vehicles, go-carts, and minibikes, are limited to a maximum of 73 dBA at 50 ft using testing methods similar to those used for EPA regulations. Any vehicle not registered for road use cannot exceed 82 dBA at 50 ft in a residential or business district.

The sounding of any horns or audible signal devices is prohibited for any purpose except as required by law. Also, boats are prohibited from generating noise levels greater than 76 dBA at 50 ft within the city or within 2 mi of the city corporate limits.

Powered Equipment

Maximum instantaneous noise levels of industrial, agricultural, or construction powered equipment (manufactured after January 1, 1982) cannot exceed 83 dBA at 50 ft and levels of residential lawn maintenance equipment (manufactured after January 1, 1978) cannot exceed 65 dBA at 50 ft. In addition, powered commercial equipment with 20 horsepower or less intended for infrequent use in residential areas such as chain saws, pavement breakers, log chippers, or powered hand tools (manufactured after January 1, 1980) cannot exceed 80 dBA at 50 ft.

Manufacturing Districts

As is the case in the New York City Zoning Resolution, different octave band noise limits are set forth for each of the three levels of manufacturing districts, M1 to M3, where M3 is the heaviest manufacturing district. The specific limits are shown in Table 5-26. Whenever an M2 or M3 district does not abut a residential, business, or commercial district, the standards listed for the M1 district are applied at the nearest residential, business, or commercial district boundary line, according to the Chicago Zoning Ordinance (Municipal Code of Chicago, Title 17). Also note that the values listed in Table 5-26 apply to the boundary of a residential, business, or commercial district for the M1 district and to the farther of either 130 ft from the closest plant property line or the boundary of a residential, business, or commercial district for the M2 and M3 districts. The values listed in Table 5-26 do not apply to construction sites.

Typical Municipal Ordinances

Typical municipal ordinances set not-to-exceed limits and consider instantaneous noise levels below 50 to 55 dBA at night and 60 to 65 dBA during the day to be acceptable. Some suburban and rural municipalities have set nighttime limits as low as 45 dBA. Many local ordinances do not quote decibel levels and only make general statements prohibiting noise in terms of nuisances, disturbances, or unnecessary noise. The definition of these

TABLE 5-26   Chicago Noise Regulation Maximum Permitted Sound Pressure Level[a]

| Octave Band (Hz) | District[b] | | | | | |
| | M1 | | M2 | | M3 | |
	Res	Bus/Com	Res	Bus/Com	Res	Bus/Com
31.5	72	79	72	79	75	80
63	71	78	71	78	74	79
125	65	72	66	73	69	74
250	57	64	60	67	64	69
500	51	58	54	61	58	63
1000	45	52	49	55	52	57
2000	39	46	44	50	47	52
4000	34	41	40	46	43	48
8000	32	39	37	43	40	45
dBA	55	62	58	64	61	66

[a] In decibels.
[b] Res, Residential; Bus/Com, business/commercial.

Source: Chicago Noise Code (1988).

terms is usually left to the governing authorities. Enforcement procedures are usually provided for sources that do not comply with the ordinance limitations.

Developing a Model

When a model does not exist for predicting noise levels caused by specific sources, such as for playgrounds, mechanical equipment, or construction noise, it is generally accepted practice to compare calculated and measured data whenever possible. An agreement between calculated and measured values of within ± 3 dBA is usually considered good enough for calibrating a model. If values vary by more than ± 3 dBA, but by a consistent amount (e.g., consistently $+5$ or -5), a correction factor is usually applied to the model that accounts for some factor that consistently causes the predicted level to disagree with measured data. For example, ground absorption or building reflections can cause sound levels to vary by more than ± 5 dBA. Therefore, the measurement environment must be examined carefully when calibrating a model. If values stray from predicted levels by more than ± 3 dBA and fluctuate broadly, the model should not be used in the analysis.

When measurement data are not available for sources in question, theoretical models can be developed; however, caution should be exercised that the theory used is generally accepted and that all environmental conditions existing are compensated for in the analysis. The model must be completely documented, including derivations, references, and assumptions with reported data.

GUIDELINES FOR DRAFTING MUNICIPAL ORDINANCES

Principal Noise Ordinance Components

Noise ordinances can take on many forms but certain components should be part of all noise ordinances to ensure that they are clearly presented and enforceable. These components include sections with purpose, definitions, sound level limitations, specific prohibited acts, exemptions, variance conditions, and enforcement procedures. The contents of these sections are described briefly below.

Purpose

This section provides an introduction to the ordinance by stating the reason the ordinance is being drafted. Usually included are statements about the negative effects of excessive noise on health, safety, and the quality of life. If the ordinance deals with only certain types of noise sources, this should be stated in this introductory section.

Definitions

This section defines the critical terms that are used in the ordinance that would need clarification to interpret the ordinance with consistency for all cases. As a minimum, area or building classification (typically in terms of residential, commercial, and manufacturing) and sound level descriptors must be clarified. Also included in this section can be sound level meter specification requirements and acceptable measurement procedures.

Sound Level Limitations

The specific sound level limits and criteria are stated in this section in terms of all applicable land uses and time restrictions. If several land use and time categories exist, it is best to present the sound level limits in tabular form. Limits can be listed in terms of absolute criteria (maximum permissible levels), relative criteria (maximum permissible increase in ambient levels caused by a particular source), or a combination of the two.

Specific Prohibited Acts

This section would include limitations, in terms of operation, scheduling, and sound levels, on activities that could take place within the boundaries of the municipality. These commonly relate to sound reproductive systems, pets, construction and demolition, loading and unloading, and idling motor vehicles. Also included in this section can be a general statement restricting any unnecessary and offending noise at any noise-sensitive location.

Exemptions

This section would include any sound sources to which this ordinance would not apply. Typical sources exempted from noise ordinances are those related to emergencies or emergency work and those that have municipal approval (e.g., parades, fairs, or celebrations). Sound associated with religious celebrations, such as church bells or chimes, are usually also exempt from these restrictions.

Variance Conditions

Variances are temporary pardons from complying with any part of the noise ordinance. They are usually considered for temporary sources, such as construction operations, or sources that would be in the best public interest. Variance proceedings are usually initiated by an application submitted by the owner of the sound source that would be or is violating the noise ordinance. The governing body would decide, based on the hardship that would result to the applicant and the public by not approving the variance weighed against the hardship the noise source would provide to the public, whether or not to grant the variance. Variances are normally granted for a limited time period, after which either the source in question must comply with the ordinance or the owner of the source must apply for another variance.

Enforcement Procedures

This section would include the formal noise ordinance enforcement procedures, including notification of violations, listing fines for specific acts, time restrictions for remedying the violation, and consequences of not remedying the violation. Also included in this section can be statements of personnel responsible for enforcement proceedings along with their training qualifications.

Recommended Criteria

Mass transit, highway, and aircraft sources are usually deferred to state and federal agency jurisdiction unless one of these sources causes public unrest. It is therefore recommended to leave these sources out of the municipal ordinance.

Most noise ordinances that use specific decibel limits use absolute maximum limits only. The use of the absolute criterion alone in noise ordinances can allow many annoying sources to comply with the enforcement standards. For example, if a noise source causes ambient levels to be 60 dBA in an area where background levels are normally 45 dBA and the noise ordinance sets a maximum permissible limit of 65 dBA, the source in question would comply with the ordinance but would have the potential of being highly annoying to residents because it causes levels to be 15 dBA higher than they are without the source. It would therefore be most practical to use a combination of relative and absolute criteria in a noise ordinance. In this case, an upper not-to-exceed decibel limit can be set. Below this limit, the amount (in decibels) that the ambient level (with the source in question operating) exceeds the background level can also be set. Because people in residential areas tend to be more sensitive to additional noise at night when they are trying to sleep, more stringent criteria should be used during nighttime hours (typically considered 10:00 P.M. to 7:00 A.M).

Many municipal ordinances use the simple criterion that a noise source need only be heard to cause a violation. This approach can cause problems because of the subjective nature of the rating method. A more logical and concrete approach to the ordinance limits would be to set clear limits in terms of sound pressure levels that comply with accepted principles of noise effects on people. Overall dBA levels are recommended over octave band levels because A-weighted levels provide a much simpler rating system that was designed to rate human reaction to sound levels. It is best to provide the simplest, most repeatable monitoring and rating procedure; the more complex the procedure becomes the more likely it is that errors will be introduced and the resulting confusion will cause a loss of enforceability and interest in the ordinance restrictions.

Given typical human reactions to changes in sound levels, the recommended criteria would be to limit offending sounds to 10 dBA above the background levels during daytime (7:00 A.M. to 10:00 P.M), hours and to 5 dBA above the background levels during nighttime (10:00 P.M. to 7:00 A.M.) hours, as long as the background levels can be recorded without the offending sound, for instantaneous (with the slow meter response speed) SPLs between 40 and 75 dBA at noise-sensitive locations. No sounds above 75 dBA, generated by specific sources (other than those considered normal components of the background noise), should be permitted in any noise-sensitive area except for those sources specifically exempted from the limitations of the ordinance. Sound levels below 40 dBA should be considered on a case-by-case basis by investigating the nature of the noise source and complaint, because complaints based on levels this low could be based on low-frequency (below 100 Hz) sources (which should be handled by a vibration ordinance) or extraordinarily sensitive circumstances. Nonrepetitive impulsive sounds should be limited to 90 dBA, using the fast response speed on the sound level meter, at noise-sensitive locations.

If background levels cannot be recorded because the offending noise source is continually operating, absolute criteria should be used, limiting noise levels on the basis of the type of environment (rural, suburban, etc.) and typical levels in the area.

A sample noise monitoring data sheet with openings for all typically required information, as discussed in Chapter 3, is included in Fig. 5-10.

At least one person in the municipal office must be designated as noise control officer (NCO). This person would have the responsibility of monitoring noise levels and enforcing the noise ordinance. To ensure proper administration, the noise control officer must be sufficiently educated in the field of noise control to make proper judgments and interpretation. If the noise control officer does not have such training, a certified noise control engineer should be retained for this purpose.

There is only one organization, the Institute of Noise Control Engineering (INCE), that certifies noise control engineers on a national level. Although engineers are normally certified through the Professional Engineer's (PE) process in the United States, there is no PE certification in noise control engineering available in most states (Oregon does offer PE certification in this field). INCE board certification involves requirements similar to PE certification, including passing a fundamentals examination, having a minimum of 8 years of field experience, and passing an 8-h professional examination. Although INCE certification may not guarantee competence in environmental noise issues, it is the only certification available in the field that requires demonstrated knowledge. A directory of INCE members can be obtained through:

Institute of Noise Control Engineering
P.O. Box 3206
Arlington Branch
Poughkeepsie, NY 12603

Included in the Appendix is a sample local noise ordinance including all key points mentioned above. Much of the format of this ordinance is based on that of model ordinances offered by the States of California and New Jersey. The ONAC also issued a model noise ordinance in 1975 (EPA, 1975), coauthored by the National Institute of Municipal Law Officers (NIMLO). The ordinance in the Appendix contains the most important and relevant information from all of these documents. The relative violation criteria are original. As a noise ordinance is a legal document that constitutes the law of a municipality, the wording of the ordinance should be reviewed by an

Reference #_____

Date/Day of Week of Measurement:
Noise Source:
Location:

Nature of Complaint (re: date, NCO):

Description of Source Area (include drawing of area on separate sheet):

Facility Operation During Measurement: Typical __ Other __ Explain:

Weather/Terrain Conditions:

Instrumentation: Manufacturer   Model #   Serial #   Certification Date
              SLM
         Calibrator
     Other -

Sound Measurements: Start/Finish Time Measurement Range (dBA) Location
          Calibration

          Background

              Ambient

Violation (circle):   Yes   No   Comments:

Reviewed and Approved by: Noise Measurements Performed by:

_____   _____
Signature   Date   Name of NCO

_____
Name/Title

FIGURE 5-10.   Noise measurement report form.

attorney before the ordinance is adopted. The model ordinance includes all technical items necessary, from an acoustic standpoint, for proper and effective enforcement.

REVIEW QUESTIONS

1. Using the information given in this chapter, write a practical noise standard including acceptable and unacceptable noise levels for a typical suburban community. Reference the appropriate documents in your derivation.
2. Considering all governmental levels, what noise regulations and standards deal with:
   a. vehicular traffic sources?
   b. rail sources?
   c. aircraft sources?
   d. stationary sources?
   e. temporary/intermittent sources?
3. According to FHWA criteria, how many PCEs would correspond to one heavy truck traveling at a speed of 65 mph?
4. An employee of a firm is exposed to the following noise doses each day.
   a. 3 h at 90 dBA, 2 h at 83 dBA, 1 h at 100 dBA, and 2 h at 67 dBA.
   b. 1 h at 110 dBA and 7 h at 65 dBA.
   would these pass an OSHA inspection? If not, name three options that could be used to ensure that they would pass, assuming the employee must spend some time in each of the environments listed.
5. A new development is planned to be built in New York City. What concerns should the developer have in terms of noise?
6. What are the most generally accepted prediction models available for traffic noise on highways and aircraft noise around airports?
7. Give three examples of identical noise sources that would be acceptable in either New York City, Los Angeles, or Chicago and not acceptable in one of the other cities.
8. List 10 possible mitigation measures for noise impacts.
9. Explain the NEPA process. What agencies use it?
10. What noise levels are considered acceptable by all agencies and organizations? What noise levels are considered unacceptable by all agencies and organizations?

References

APTA. 1981. *Guidelines for Design of Rapid Transit Facilities.* Washington, D.C.: American Public Transit Association, Section 2-7.

Barry, T.M. and J.A. Reagan. 1978. *FHWA Highway Traffic Noise Prediction Model.* FHWA-RD-77-108, Washington, D.C.: U.S. Department of Transportation Federal Highway Administration Office of Environmental Policy.

Cervantes, R.W., J.M. Fergeson, and A.A. Rivasplata. 1990. *State of California General Plan Guidelines.* Sacramento, CA: State of California Office of Planning and Research.

EPA. 1974. *Information on Levels of Environmental Noise Requisite to Protect Public Health and Welfare with an Adequate Margin of Safety.* EPA 550/9-74-004, Washington, D.C.: U.S. Environmental Protection Agency Office of Noise Abatement and Control.

EPA. 1975. *Model Community Noise Control Ordinance.* EPA 550/9-76-003, Washington, D.C.: U.S. Environmental Protection Agency Office of Noise Abatement and Control.

EPA. 1982. *Guidelines for Noise Impact Analysis.* EPA 550/9-82-105, Washington, D.C.: U.S. Environmental Protection Agency Office of Noise Abatement and Control.

FAA. 1989. *Airport Noise Compatibility Planning.* FAR Part 150, Washington, D.C.: U.S. Department of Transportation Federal Aviation Administration.

FHWA. 1991. *Summary of State Highway Agency Noise Policy Definitions.* Washington, D.C.: U.S. Department of Transportation Federal Highway Administration Office of Environment and Planning.

FICON. 1992. *Federal Agency Review of Selected Airport Noise Analysis Issues.* Washington, D.C.: Federal Interagency Committee on Noise.

GSA. 1987. *GSA Supplement to Masterspec.* Section 01040. Washington, D.C.: General Services Administration.

Hanson, C.E. et al. 1993. *Guidance Manual for Transit Noise and Vibration Impact Assessment.* UMTA-DC-08-9091-90-1. Washington, D.C.: U.S. Department of Transportation Federal Transit Administration.

HUD. 1985. *The Noise Guidebook.* HUD-953-CPD. Washington, D.C.: U.S. Department of Housing and Urban Development.

PennDOT. 1987. *Environmental Impact Assessment and Related Procedures.* Design Manual, Chapter 8, Part 1A, Harrisburg, PA: Commonwealth of Pennsylvania Department of Transportation.

Shapiro, S.A. 1991. *The Dormant Noise Control Act and Options to Abate Noise Pollution.* Washington, D.C.: Administrative Conference of the United States.

USAF. 1992. *AICUZ Handbook, A Guidance Document for Air Installation Compatible Use Zone (AICUZ) Program.* Washington, D.C.: U.S. Air Force.

VA. 1988. *Construction and Valuation Policies, Procedures, and Methods, Loan Guaranty Operations for Regional Offices.* DVB Manual M26-2, Chapter 2, Section VIII. Washington, D.C.: Veterans Administration Department of Veterans Benefits.

6

Common Noise Sources

Noise sources encountered in everyday life are discussed in this chapter in terms of their measured levels, frequency content, and potential for damaging the human hearing mechanism. The sources are divided into six general categories: recreational, transportation, public maintenance, household, natural and animal, and sound reproductive sources.

Before this discussion begins, it is appropriate to review acceptable noise limits that are used on a standard basis. Compatible land use tables for regulatory agencies were given in Chapter 5. American National Standards Institute (ANSI) Standard S12.40-1990, *American National Standard Sound Level Descriptors for Determination of Compatible Land Use*, provides a land use compatibility table in terms of annual average (yearly) L_{dn} values. Table 6-1 summarizes this table.

These limits are generally accepted for annoyance purposes. In terms of hazardous limits, Occupational Safety and Health Administration (OSHA) 8-h and Environmental Protection Agency (EPA) 24-h L_{eq} limits of 85 and 70 dBA, respectively, are generally regarded as initial levels of concern for noise-induced hearing loss resulting from long-term (20 to 40 years) exposures. The lower limit for acoustic trauma, having the potential to cause immediate permanent hearing damage, is the OSHA instantaneous peak limit of 140 dB.

It must be noted that noise levels below the acoustic trauma threshold normally do not cause immediate permanent hearing loss. The amount of hearing loss attributable to such sources depends on the frequency and duration of the exposure. Of course, higher levels require shorter durations and frequencies of exposure to do damage than do lower levels.

For general reference, if conversation is impossible to hear in a certain area, the noise levels are probably higher than 90 dBA.

226

TABLE 6-1   General Land Use Compatibility Guidelines[a]

Land Use Category	Compatible	Marginally Compatible	Incompatible
Residential			
Single family, much outdoor use	<55	55–65	>65
Multifamily, moderate outdoor use	<60	60–65	>65
Multistory, limited outdoor use	<60	60–65	>65[b]
Transient lodging	<65	65–70	>70[c]
Schools, libraries, churches, hospitals, nursing homes	<60	60–65	>65[b]
Auditoriums, concert halls	<60	60–65	>65
Music shells	No rating	50–65	>65
Sports arena, outdoor spectator sports	<60	60–70	>70
Neighborhood parks	<55	55–70	>70
Playgrounds, golf courses, riding stables, water recreational facilities, cemeteries	<60	60–75	>75
Office buildings, commercial and professional, livestock farming, animal breeding	<65	65–75	>75
Wholesale, industrial, manufacturing, utilities	<70	70–80	>80
Agricultural (except livestock)	<75	75–85	>85
Extensive natural wildlife and recreation	<60	60–75	>75

[a] In dBA L_{dn}.

[b] Compatible for 65–75 with insulation to provide an interior L_{dn} of 45 dBA.

[c] Compatible for 70–75 with insulation to provide an interior L_{dn} of 45 dBA.

Source: ANSI S12.40-1990, Appendix. (Reprinted by permission of the Acoustical Society of America, New York, NY.)

RECREATIONAL SOURCES

Most of the sounds we are exposed to in our daily lives do not pose any threat to our comfort or hearing abilities. There are some sources, however, that can be hazardous. The most common recreational activities that have potential for causing hearing loss include shooting firearms (in target shooting or hunting) and listening to amplified music (either from personal or public systems). Amplified music is discussed in Sound Reproductive Systems (below).

Many studies have been performed relating recreational gun use to hearing loss. Studies performed over the past 30 years state measured peak sound pressure level (SPL) values from commonly used pistols, rifles, and shotguns ranging from 132 to 172 dBA (Clark, 1991). These are in the range of acoustic trauma potential, where immediate permanent hearing loss can result from a single exposure. Recreational gun use is commonly regarded as one of our most serious nonoccupational noise hazards. More than 50%

of the men in the industrial work force fire guns at least occasionally and a single shot from a typical firearm has the acoustic energy equivalent of a 1-week, 8-h daily exposure of 90 dBA. When we account for the 50 rounds of ammunition in a typical box of shells, firing one box of shells can be equated to an 8-h daily exposure of 90 dBA for 1 year (Clark, 1992). Another study in this area went even further to conclude that occasional exposure to recreational gunfire can be considered equivalent to an 8-h daily exposure of 89 dBA for 20 years (Johnson and Riffle, 1982). In all cases, hearing loss was an obvious result of exposures to recreational gun use. It is estimated that 50 million Americans own and use firearms and most hunters do not wear hearing protection (Cherow, 1991).

Other common recreational noise sources that have been documented to generate noise levels in the range producing acoustic trauma include fireworks and cap guns.

Any number of recreational activities can cause community annoyance issues. On a mass scale, activities at amphitheaters, stadiums, amusement parks, and automobile racing facilities (to name a few) can generate intrusive noise levels thousands of feet from their sources. Careful planning in facility design and community control through drafting effective noise ordinances are the most effective ways of dealing with such problems.

TRANSPORTATION SOURCES

Transportation sources are the greatest contributors to the outdoor ambient noise environment. These sources include automobiles, motorcycles, trucks, buses, trains, and aircraft. Of all transportation sources, surface vehicles (cars, trucks, and buses) are the greatest contributors to the overall noise environment. The EPA performed an extensive survey of noise levels throughout the country and concluded that 87% of urban populations are exposed to L_{dn} levels exceeding the EPA impact criteria for residential uses of 55 dBA (EPA, 1982). They noticed no significant variation in noise levels on weekends or other times.

Regulations limiting exterior noise levels of these sources are discussed in Chapter 5. However, people are exposed not only to noise levels exterior to these sources but also inside these sources when they are passengers. Both exterior and interior noise measurements are summarized below for typical transportation sources. Average sustained (nonimpulsive) noise levels for passengers on typical models of these sources are listed in Table 6-2 and maximum sustained noise levels outside of these vehicles at distances of common exposures are listed in Table 6-3.

All levels listed in Table 6-2 are under typical moving conditions except for automobiles as noted. It must be noted that interior levels of surface

TABLE 6-2   Noise Levels Inside Common
Daily Transportation Sources

Vehicle	Interior Noise Level (dBA)
Automobile, 55 mph	69–78
Train	
Amtrak	63–67
Commuter	69–73
Subway	74–79
Airplane	
Long range	70–80
Commuter	75–85
Helicopter	85–95

Sources: Cowan (1992); Thompson et al. (1991).

vehicles are highly dependent on the road surface. Variations in interior levels by as much as 15 dBA can occur between driving on smooth tar and rough concrete road surfaces.

Aircraft flyovers are normally rated in terms of sound exposure level (SEL). Table 6-4 lists SEL values for common airplanes and helicopters. For reference, maximum noise levels are typically 5 to 10 dBA lower than the SEL values.

PUBLIC MAINTENANCE SOURCES

Public maintenance sources include construction equipment and machinery that is used for maintenance in public areas. Table 6-5 lists noise levels for typical public maintenance sources.

TABLE 6-3   Noise Levels Outside Common
Daily Transportation Sources
at Full Speed

Vehicle	Exterior Noise Level (dBA at 30 ft)
Automobile	72–75
Bus	82–87
Freight train	85–88
Subway train	98–103
Truck	82–89
Truck, idling	70–75

Source: Cowan (1992).

TABLE 6-4   Noise Levels of Common Aircraft

Aircraft Model	SEL (dB at 1000 ft)
Airplanes	
707, DC-8	113.5
727	112.5
737, DC-9	110.0
747	102.5
DC-10, L-1011	100.0
Learjet	97.1
757	97.0
767	96.7
Helicopters	
Boeing CH-47C	99.2
Sikorsky S-64	93.2
Augusta A109	89.7
Sikorsky CH-53	89.1
Bell 212	89.0
Sikorsky S-61	86.4
Sikorsky S76	85.8
Bell 206L	79.4

Sources: FICON (1992); Newman et al. (1982).

TABLE 6-5   Maximum Noise Levels of Common Public Maintenance Sources

Noise Source	Noise Level (dBA)
Augered earth drill	94 at 10 ft
Backhoe	87–99 at 30 ft
Backhoe, idling	74 at 30 ft
Cement mixer	77–85 at 10 ft
Chain saw cutting trees	89–95 at 10 ft
Circular saw on concrete	91 at 30 ft
Compressor, street	91 at 3 ft
Front-end loader	79–93 at 50 ft
Garbage truck	85–97 at 10 ft
Jackhammer	100 at 3 ft, 96 at 10 ft
Paving breaker	86 at 30 ft
Steamroller	87 at 30 ft
Street cleaner	74 at 30 ft
Street paver	84 at 30 ft
Wood chipper	
Tree shredding	93 at 30 ft, up to 101
Idling	85 at 30 ft

Source: Cowan (1992).

Other sources that can be included in this category are industrial facilities such as power plants and other utility-related facilities. These are typically governed by utility regulatory agencies that may or may not have noise standards. As long as the source is not regulated by a federal or state agency, an effective local ordinance can be drafted that can be used to resolve noise-related disputes.

HOUSEHOLD SOURCES

Common household sources include appliances and equipment encountered both inside and outside the home, normally used for private maintenance. These would include such equipment as vacuum cleaners, hair dryers, and lawn mowers. Maximum sustained noise levels at typical distances between the user's ear and the source for household sources are listed in Table 6-6. Sources listed are those that commonly generate the highest levels of sound. Office equipment manufacturers for home-based offices have produced products that are quiet enough not to be included in this list.

NATURAL AND ANIMAL SOURCES

Natural Sounds

Natural sources include such things as wind, water, insects, and non-domesticated animals. Maximum sustained noise levels for typical natural sources are listed in Table 6-7.

Background noise levels were monitored recently in several national parks (Bommer and Bruce, 1992; Grasser and Moss, 1992). These readings quantify some of the quietest outdoor areas around with $L_{90(1)}$ values down to 17 dBA and below 10 dB in octave band frequencies between 250 and 2000 Hz. Outdoor sound levels this low would be indicative of no natural activity. As is shown in Table 6-7, noise generated from wind, water, animals, or insects would cause ambient levels to increase significantly above these minimal levels. For reference, very quiet suburban areas during the winter at night have ambient levels in the 30- to 40-dBA range.

Animal Sounds

Animal sources include domesticated animals such as dogs and cats. Dog barking can be quite intrusive in residential neighborhoods, even when the dog is inside. Single dogs can typically generate more than 90 dBA at 5 ft. Their dominant barking frequency range also lies in our most sensitive zone. In dog kennels where many dogs can be barking in unison, maximum

TABLE 6-6   Noise Levels of Common Household Sources

Noise Source	Noise Level (dBA)
Dehumidifier	58–60 at 5 ft
Food blender	76–81 at 3 ft
Garbage disposal	76–78 at 3 ft
Microwave oven	56–58 at 3 ft
Sink faucet, full power	71–73 at 3 ft
Hand-held vacuum cleaner	82–87 at 3 ft
Small rechargeable vacuum	75–77 at 3 ft
Vacuum cleaner	78–85 at 5 ft
Steamer carpet cleaner	84–92 at 5 ft
Portable hair dryer	77–86 at 1 ft
Lawn edger	89–93 at 5 ft
Leaf blower	87–93 at 5 ft
Power lawn mower	
Hand	81–86 at 5 ft
Riding	88–93 at 5 ft

Source: Cowan (1992).

readings can exceed 100 dBA. As can be testified by many homeowners, dog barking can be extremely annoying because of its impulsive nature, clear audibility over most background levels both indoors and outdoors, and frequency content in the most annoying frequency range.

Cats typically generate levels much lower than dogs, in the 60- to 70-dBA range. They usually cause annoyance only when outdoors in quieter areas, but their dominant voice frequency range is also in the most annoying range for humans.

TABLE 6-7   Noise Levels of Common Natural Sound Sources

Noise Source	Noise Level (dBA)
Rustling leaves in wind	55–58, up to 66
Typical mall fountain	72–74 at 10 ft
Medium-size waterfall	69–70 at 10 ft
Ducks, geese	
Single	63–68 at 30 ft
Group	71–75 at 30 ft
Holstein cow	Up to 94 at 10 ft
Summer nighttime insects	50–54 in open field

Source: Cowan (1992).

TABLE 6-8   Noise Levels of Crowds of People

| Event | Crowd Size | Crowd Noise Level (dBA) | | |
		L_{eq}	Maximum	Event Time
Playground recess	100–500	68–77	101	15–30 min
Basketball game	12,600	89	107	2.5 hr
Ice hockey game	17,400	90	113	3 hr
Wrestling match	7,000	89	111	2.5 hr
Football game	65,000	88	111	3 hr
Rock concert	19,000	NA[a]	109	3 hr

[a] L_{eq} was dominated by concert noise.

Sources: Cowan (1992); NYCSCA (1992) (with permission).

Also included in this category would be the unamplified human voice, which can generate significant noise levels when individual voices are raised or in localized groups in playgrounds or at spectator events. The typical human speaking voice at 5 ft is in the 60- to 70-dBA range. An individual person yelling can generate 100 to 103 dBA at 5 ft, independent of gender and age. When in large groups, these numbers can increase to as much as 110 dBA or more. The most common large gatherings of people occur in school playgrounds and at spectator events.

Crowd noise levels were monitored recently at eight school playgrounds in the New York City area and at a sampling of spectator events (NYCSCA, 1992; Cowan, 1992). The spectator events included the professional sports of baseball, football, wrestling, basketball, tennis, and hockey, and a sold-out rock concert. In all cases, instantaneous crowd noise levels exceeded 90 dBA. Table 6-8 lists L_{eq} and maximum levels measured for a sampling of events where crowd noise clearly dominated the noise environment. Note that the basketball, hockey, wrestling, and concert crowds were recorded in the same indoor arena. The football crowd was recorded in a stadium open at the top.

SOUND REPRODUCTIVE SYSTEMS

A sound reproductive system is a device that artificially reproduces and amplifies a sound signal. The device that performs this reproduction is known as a transducer, simply defined as a device that transforms energy from one form into another. Our hearing mechanism, described in Chapter 1, is an example of an acoustic transducer that converts acoustic energy into mechanical energy in the middle ear by causing the ossicles to vibrate and transforms mechanical energy into electrical energy as the inner ear hair cells send electrical signals to the auditory nerve that are generated in response to the mechanical movement of the hair cells.

Man-made sound system components are also acoustic transducers. The loudspeaker is a transducer that transforms electrical energy into acoustic energy. Although there are many shapes and sizes of loudspeakers available, their basic designs are similar to each other. This is discussed below, followed by a discussion of the three general categories of loudspeaker systems that we are exposed to most often: personal systems, public address systems, and concert systems.

Basic Design

The most common type of loudspeaker is the dynamic, or moving coil, design. Figure 6-1 shows a cross-sectional view of the circular loudspeaker sound source. The physical principle of operation behind this design is that a wire (known as the voice coil) with electric current running through it and in the presence of a magnetic field induces a force proportional to the magnetic field strength, the voice coil length, and the current. The current running through the voice coil varies in proportion to the sound signal that is being reproduced. The force resulting from this interaction sets the voice coil into vibration, which in turn vibrates the attached coil former and diaphragm. The vibrating diaphragm then radiates sound to the outside medium (usually air).

When a loudspeaker such as the one described above radiates sound directly to the air, acoustic energy is being transferred inefficiently because of the large acoustic impedance mismatch between the loudspeaker and the

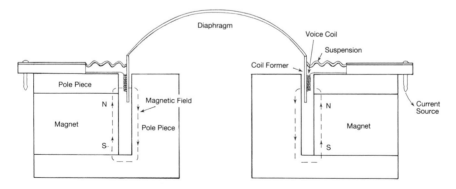

FIGURE 6-1.   Cross-sectional view of a dynamic loudspeaker.

outside air, as is discussed in the section on hearing in Chapter 1. This impedance mismatch is caused by the sharp change in the cross-sectional area in the sound propagation path, as would also be the case for the human ear if we did not have pinnas to provide a smoother cross-sectional area transition for the sound to pass through. In the case of loudspeakers, the solution to this problem is the horn.

A horn is a device with solid walls that surround the loudspeaker diaphragm and gradually flare out to provide a more gradual cross-sectional area change, and thus a smoother acoustic impedance transition, between the loudspeaker diaphragm and the outside environment. The rate at which this flare occurs is described by mathematical functions and variables depending on the frequency range of sound reproduction desired. Larger horns typically reproduce lower frequencies more efficiently because their associated wavelengths are larger. The end of a horn closest to the diaphragm is called the throat and the end open to the outside is known as the mouth.

Size and weatherproofing considerations caused loudspeaker designers to develop a folded type of horn design, its cross-section shown in Fig. 6-2. The air column length is the same in each horn shown but the folded design provides the user with compactness and weatherproof speaker protection. This type of horn design is used extensively in outdoor and industrial environments. Because size restrictions limit lower frequency sound reproduction for horns in general, folded horns are best suited for voice paging and warning systems.

A principal characteristic of sound radiation patterns from horns is a narrowing, or beaming as it is commonly known, of the coverage angle (the angle, with respect to the loudspeaker, over which the best frequency response exists) as frequency increases. Coverage angles are usually derived from response curves such as those shown in Fig. 6-3. As is shown, the coverage angle of a loudspeaker is defined as the angle between locations at which the coverage sensitivity drops off by 6 dB.

Because speech intelligibility is dependent on consonant (higher frequency) recognition and many loudspeaker manufacturers specify coverage angles at frequencies less than 2000 Hz, the useful coverage angle for traditional horns is often less than that quoted in product literature. A plot of coverage angle versus frequency is usually necessary to estimate the horn coverage for different applications. Because many manufacturers do not provide such information, at least a coverage angle at 2000 Hz should be stated for proper estimation of intelligible speech coverage.

The search for a practical alternative to horn beaming has resulted in the so-called constant directivity horn. If designed properly, this type of horn produces a constant coverage angle within a desired frequency range. In this case, a graph of coverage angle versus frequency should approximate a

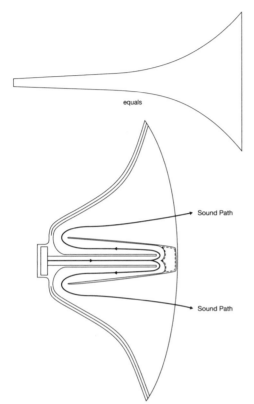

equals

Sound Path

Sound Path

FIGURE 6-2.   The folded horn design provides the same sound path length as a long horn, but in a smaller size.

horizontal line (within a ± 5 dB tolerance) over the desired frequency range at the coverage angle quoted. A horn shape that generates a constant directivity coverage pattern is shown in Fig. 6-4. Originally designed by D. Keele at Electro-Voice, Inc. (Buchanan, MI) in 1975, it consists of a narrow, nearly parallel-walled channel throat feeding into a straight, angled (conical) section and flaring out over the outer third of the horn length in a curved mouth section. This outer section shape can be an arc of a circle or a curve based on the horn length dimension raised to a power (exponential).

Most loudspeakers used for commercial purposes are based on the same design of Fig. 6-1 and are coupled to paper (or some light, thin material) cones. This increases the frequency response capability of the unit but makes it much more fragile than industrial loudspeakers connected to metal or

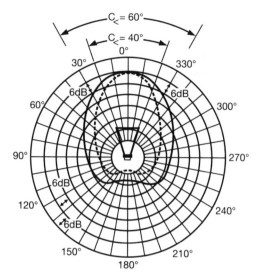

FIGURE 6-3. Directivity response curves for coverage angles ($C_<$) of 40°and 60°.

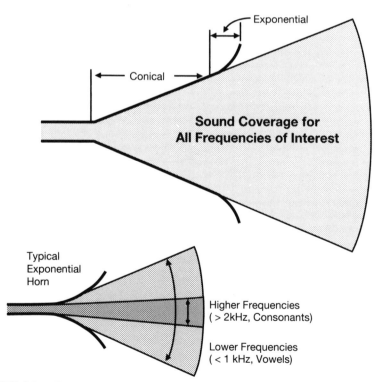

FIGURE 6-4. Coverage patterns of constant directivity vs. typical horns.

plastic horns. The frequency response of a cone loudspeaker can be significantly improved by mounting it in a wall or baffle. Figure 6-5 illustrates how different loudspeaker mounting techniques improve loudspeaker response by reducing sound cancellations between the front and back of the radiating cone surface. Recalling that decreases in frequency translate to increases in wavelength size, larger baffles mean lower frequency cancellation and smooth (output equals input) frequency response to lower frequency limits.

A closed box essentially becomes an infinite baffle and minimal cancella-

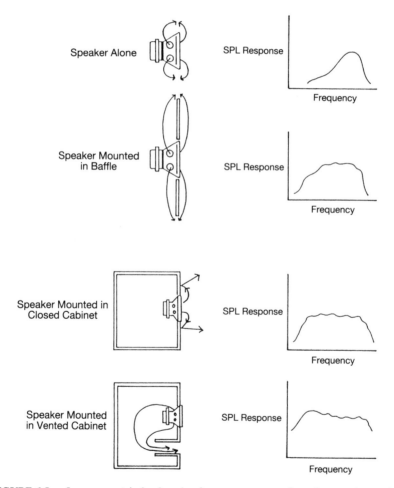

FIGURE 6-5.   Improvement in loudspeaker frequency response through mounting method.

tion should occur; however, we do not hear lower frequencies with the same sensitivity as higher frequencies. The resultant sound has minimal perceived bass because, for all frequency components to sound equal in level, the low frequencies would have to be artificially boosted in level. This can be accomplished by cutting a hole in the loudspeaker cabinet to create a vent that sets up a resonance inside the box at the resonance frequency dictated by the dimensions of the box (similar to the Helmholtz resonator described for its absorption capabilities in Chapter 4). The sound that sets up this resonance in the box is generated by the rear side of the loudspeaker cone. The vent, or port, then becomes a new radiator of a narrow frequency range of amplified sound centered around the resonance frequency of the inside of the cabinet–port combination. This is a common design of home and professional loudspeakers.

Because each loudspeaker component, because of its physical dimensions compared with that of the reproduced sound wavelengths, best reproduces a limited acoustic frequency range, high-fidelity loudspeakers incorporate several speakers of different size into one cabinet. These are commonly known as the woofer (largest), covering the lowest frequencies; the mid range, covering the speech frequency range; and the tweeter (smallest), covering the highest frequencies. Also available are subwoofers for those who want sounds below 100 Hz to be stressed, usually built into large cabinets, and super-tweeters to extend the high-frequency response of the loudspeaker system.

Smaller loudspeakers are currently being designed that retain low-frequency response that would not be expected from units of their size. The key to their low-frequency response is a long sound path that is built inside the loudspeaker cabinets by folding the path several times into a labyrinth.

In addition to dynamic loudspeaker designs, electrostatic and piezoelectric designs are also common. These involve charged plates that cause materials to vibrate in sympathy with the incoming electrical signal and thus generate sound.

If one is in the market to buy a set of loudspeakers, it should be noted that the tolerances of many loudspeaker specifications, such as total harmonic distortion, are beyond the detection limits of the human hearing mechanism. Because taste in sound is subjective, the best advice for choosing the best loudspeakers to purchase is to listen to the speakers available and compare their sound in terms of each individual's acoustic palate. Always beware the sound rooms set up in audio stores that attempt to set up an ideal listening environment. The loudspeakers are usually set up and located to sound their best in this room that has room effects unlike those in most residential environments. A common complaint is that the new loudspeakers installed in the house do not sound as good as they did in the showroom, and they never will unless there is an acoustic environment similar to that of the showroom in the listening area of the house. It should come as no surprise that the most expensive loudspeakers are usually displayed in the sound rooms of these stores. In many cases, a $200 pair of loudspeakers may sound better to a listener than a $2000 pair. Price does not necessarily dictate the best loudspeaker for each person. The sound should.

Personal Systems

Personal sound reproductive systems can be divided into two categories, depending on their portability. Nonportable systems are the type that are installed in typical households having separate amplifiers, tuners, some type of sound reading equipment (e.g., a compact disc, cassette, or record player), and loudspeakers. These are typically capable of generating SPLs exceeding 100 dBA; however, neighborly concerns and noise ordinances normally restrict these emissions both outdoors and indoors because of the transmission loss (TL) limits of common building materials (as specified in the building codes discussed in Chapter 7).

Portable systems have all components in one package that are also capable of generating SPLs above 100 dBA. "Boombox" types of systems have limited hazard potential for the same reasons nonportable systems are usually not a problem. However, because they are often used in public, their clear annoyance potential need not be elaborated on. The type of portable system having headphones has the greatest hazard potential of any personal systems. These systems set the loudspeakers at the entrance to the ear canal with maximum energy directed into the ear and minimum energy directed away from the ear. Even though the loudspeaker components are small, a broad frequency range can be reproduced because lower frequencies can be sensed when the speaker is close to the ear. At any distance away from the speaker, only the high frequencies characteristic of such small speakers can be heard.

Using these systems, SPLs can be raised to levels at the subject's ear (without endangering or annoying the public) that would not be acceptable to others using the other types of personal systems. Although maximum SPLs from these systems have been reported up to 128 dBA, most systems generate SPLs between 60 and 114 dBA (Clark, 1991). This is one case where the potential hazard of the source is determined by the user of the equipment. It has been suggested by many that warning labels be placed on these systems designating the potentially hazardous range of levels (Cherow, 1991).

Public Address Systems

Public address systems can be divided into two categories according to their intended use, for voice or warning signal reproduction.

Voice Systems

When indoors, any type of loudspeaker can be used that reproduces sound reliably over the speech frequency range. Most commercial loudspeakers are appropriate for these situations. When high levels of sound are required or when loudspeakers must be located outdoors or in industrial environments, horns with dynamic compression drivers are most commonly used. Folded horns are typically used to protect the loudspeaker diaphragm from damage. To be clearly intelligible, the system must generate SPLs at least 10 dBA above the background noise levels. Although these systems are typically capable of generating SPLs above 100 dBA, sounds begin to distort in our hearing mechanisms at these levels. Considering that background levels would be above 90 dBA to require voice systems above 100 dBA, people should be wearing hearing protection in such cases. Because typical hearing protection devices (HPDs) lower the background and signal of interest by the same amount, speech intelligibility for voice systems can be intelligible at such high levels when wearing HPDs.

Warning Systems

Warning systems can be in terms of voice or tones. It is often easier to identify tones than to decipher speech, especially in environments with background levels above 85 dBA. A tone need only be 3 dB higher than the background level to be clearly audible. Because all sound energy is concentrated into one small frequency band, more sound energy can be generated for tones than for voice. Tonal warning systems are available that generate 125 dBA 100 ft from the source. Distortion is also not a problem with tones because it is important only to hear the tone rather than to interpret it.

Tonal warning systems are available in terms of compression driver–loudspeaker combinations or sirens. The advantage of the loudspeaker is that it can be aimed at specific areas. Typical sirens radiate noise in all directions (omnidirectionally). Most sirens used on fire stations, ambulances, and fire trucks can generate SPLs of more than 105 dBA at 50 ft. Sirens on fire stations need to be that loud in many cases to alert firemen over a large geographic area; however, sirens on vehicles need not be that loud to alert people and other vehicles in their path.

Concert Systems

Large Rock Concert Systems

Until recently, loudspeaker setups for rock concerts typically used stacks of loudspeakers on each side of the stage facing out into the audience. Using this setup, the people sitting closest to the stage received the highest noise exposures and levels decreased with distance away from the stage. In these cases, levels exceeding acoustic trauma limits were common for audience members close to the loudspeakers. The performers that remained behind the loudspeaker towers were usually not at serious risk for hearing damage because their exposures were limited to the noise emitted from the backs of the loudspeakers and the noise reflected back to them from arena walls.

More common designs now include systems of loudspeakers hanging from supports (known as flown speakers) above and around the stage. The loudspeakers are set in a pattern such that each speaker, or group of speakers, faces different sections of the audience. Each speaker facing a specific section can be adjusted separately to provide similar sound levels everywhere in the audience area. Although this flown design provides some relief from acoustic trauma to the audience members close to the stage, everyone in the arena, including the performers, is now routinely exposed to SPLs exceeding 110 dBA at rock concerts.

Many studies have been performed on sound levels at rock concerts over the past 20 years. In summarizing these studies, Clark (1991) reported a geometric mean exposure level of 103 dBA, with individual studies reporting up to 120 dBA. A single study was performed on a Bruce Springsteen concert in St. Louis in 1986 (Clark and Bohne, 1986) for which average exposure levels of 100 to 101 dBA were reported. Interestingly, a recent study on a Bruce Springsteen concert performed in Philadelphia (Cowan, 1992) resulted in the identical average exposure levels of 101 dBA. This is probably not a coincidence because the touring crew can set concert levels the same in different arenas. Maximum instantaneous SPLs recorded at the concert overloaded the monitoring instrument, which has an upper peak SPL limit of 149 dBA. These highest levels were recorded at the farthest location from the stage, showing that any seat at a concert can be hazardous.

Accompanying many of the studies mentioned above were evaluations of noise-induced hearing loss for audience members. A temporary threshold shift accompanied by tinnitus was a common result of these concert exposures, but nearly all subjects regained full hearing ability within a few days of the concerts. The positive aspect of noise exposure at rock concerts is that most people attend them infrequently. Although noise levels at rock concerts are very high, they seem to be dangerous only to those that frequently attend the events.

Club Systems

Sound levels in clubs where bands or sound systems play music also routinely exceed 100 dBA, but are usually not as high as those at most rock concerts. There is no typical loudspeaker setup for these venues and hearing loss susceptibility depends on the frequency and duration of attendance.

Sound Bleeding to Noise-Sensitive Areas

A significant environmental noise issue is the noise from concert and club systems carrying into noise-sensitive areas. The high levels of sound generated in concert spaces can bleed into surrounding areas to disturb neighbors because of the limitations of the buildings to contain the sound completely. Outdoor amphitheaters do not even provide side building walls to reduce sound transmission and, in many cases, concert sounds can be heard more than 1000 ft from the facility. Low-frequency (below 300 Hz) sound tends to be the biggest problem among all types of facilities. These sources are typically dealt with through local ordinances. It is important that such issues are handled properly in each ordinance to make it enforceable.

The most practical way to reduce exterior noise exposure in existing facilities, where noise control measures have not been incorporated into the facility design, is to increase the number of loudspeakers in the facility such that each speaker (or group of speakers) is closer to the audience members and can therefore be set to emit a lower level of sound. This distributed layout of loudspeakers can significantly reduce exterior sound transmission with minimal compromising of concert atmosphere.

References

Bommer, A.S. and R.D. Bruce. 1992. Long-term ambient sound monitoring in national parks. Sound and Vibration, February 1992, 16–18.

Cherow, E. 1991. *Combatting Noise in the '90s: A National Strategy for the United States.* Rockville, MD: American Speech–Language–Hearing Association.

Clark, W.W. 1991. Noise exposure from leisure activities: a review. J. Acoust. Soc. Am. 90(1):175–181.

Clark, W.W. 1992. Hearing: the effects of noise. Otolaryngol. Head Neck Surg. 106(6):669–676.

Clark, W.W. and B.A. Bohne. 1986. Temporary threshold shifts from attendance at a rock concert. J. Acoust. Soc. Am. 79(Suppl. 1):S48–S49.

Cowan, J.P. 1992. Personal noise monitoring data.

EPA. 1982. *National Ambient Noise Survey.* EPA 550/9-82-410. Washington, D.C.: U.S. Environmental Protection Agency Office of Noise Abatement and Control.

FICON. 1992. *Federal Agency Review of Selected Airport Noise Analysis Issues.* Washington, D.C.: Federal Interagency Committee on Noise.

Grasser, M.A. and K. Moss. 1992. The sounds of silence. Sound and Vibration February 1992, 24–26.

Johnson, D.L. and C. Riffle. 1982. Effects of gunfire on hearing level for selected individuals of the Inter-Industry Noise Study. J. Acoust. Soc. Am. 72(4):1311–1314.

Newman, J.S., E.J. Rickley, and T.L. Bland. 1982. *Helicopter Noise Curves for Use in Environmental Impact Assessment.* FAA-EE-82-16. Washington, D.C.: U.S. Department of Transportation Federal Aviation Administration.

NYCSCA. 1992. SCA Playground Noise Study. New York: New York City School Construction Authority.

Thompson, P. et al. 1991. *Civil Tiltrotor Missions and Applications Phase II: The Commercial Passenger Market.* NASA CR 177576. Moffett Field, CA: National Aeronautics and Space Administration.

7

Construction Issues

BUILDING CODES

The building codes most commonly used in the United States are the BOCA, UBC, and Standard Building Codes. In terms of noise, these codes deal only with the required sound attenuation between dwelling units of multifamily dwelling unit buildings. In addition, the General Services Administration (GSA) has standards that are used for designs of public buildings. The noise criteria of these codes and standards are discussed below, along with the California Building Code as an example of a state's requirements.

BOCA

The BOCA National Building Code is published by the Building Officials and Code Administrators International, Inc. (Country Club Hills, IL). According to the eleventh edition, 1990 version, Section 714, a minimum sound transmission class (STC) of 45 is required for all walls and floor–ceiling assemblies between dwelling units or between a dwelling unit and a public area. A minimum impact insulation class (IIC) of 45 is also required for floor–ceiling assemblies. References are made to three documents containing accepted ratings for specific assemblies (BIA, 1988; GA, 1992; NCMA, 1990).

UBC

The Uniform Building Code (UBC) is published by the International Congress of Building Officials (Whittier, CA). According to the 1991 edition, Appendix Chapter 35, a minimum STC of 50 is required for all walls and floor–ceiling assemblies between dwelling units or between a dwelling unit and a public area. A minimum IIC of 50 is also required for floor–ceiling assemblies. If the STC and/or IIC must be field tested, minimum field sound transmission class (FSTC) and field impact insulation class (FIIC) values of 45 must be measured. In addition, doors leading outside dwellings of multidwelling unit buildings must have a minimum rated STC of 26. Reference is made to the Gypsum Association *Fire Resistance Design Manual* (GA, 1992) for acceptable designs.

Standard Building Code

The Standard Building Code is published by the Southern Building Code Congress International, Inc. (SBCCI) (Birmingham, AL). According to the 1991 edition, Appendix K, a minimum STC of 45 is required for all walls between dwelling units or between a dwelling unit and a public area. References are made to the same documents that the BOCA National Building Code mentions (BIA, 1988; GA, 1992; NCMA, 1990) for acceptable designs.

In addition to the building code, the SBCCI published the *SBCCI Standard for Sound Control* in 1987, providing a model noise ordinance similar to those discussed in Chapter 5 of this book.

General Services Administration

The GSA has published standards for acoustic insulation for rooms in public court buildings. These standards are part of the PBS PQ100 document, published in September 1992. They provide much more detailed requirements than typical building codes because of the more extensive soundproofing requirements for courtrooms. In terms of insulation between rooms, ratings are listed in terms of STC and noise isolation class (NIC) values, ranging from 40 for general office spaces to 65 for courtrooms. Most private rooms require minimum NIC values of 45, with higher values required for extrasensitive spaces.

California State Building Code

California has its own state building code that is in Part 2, Title 24 of the California Code of Regulations, originally published in 1974 and current as of December 1988 (according to the California Department of Health Services). Appendix Chapter 35 requires a minimum STC of 50 for all walls and floor–ceiling assemblies between dwelling units or between a dwelling unit and a public area. A minimum IIC of 50 is also required for floor–ceiling assemblies. If the STC and/or IIC must be field tested, minimum values of 45 must be measured. In addition, doors leading outside dwellings of multidwelling unit buildings must have a minimum rated STC of 26.

Reference is made to the Gypsum Association manual (GA, 1992) and the ratings handbook published by the California Department of Health Services (DuPree, 1988) for acceptable designs.

In addition to the above similarities to the UBC, the interior noise level limit (attributable to outdoor sources) is 45 dBA L_{dn} or community noise equivalent level (CNEL), the L_{dn} being the preferred descriptor.

COMMON NOISE PROBLEMS AND SOLUTIONS

There are often expectations of privacy in offices and dwellings that are not realized because of construction methods that eliminate any possibility of acoustic privacy. Often people move into new offices and housing units and find that they can easily hear people and activities in other rooms or dwellings. This is not only a source of annoyance that can affect job performance or restful sleep, but it also adds stress to people knowing that all they do and say can be heard by others in other rooms. This is of particular annoyance when the developer or realtor has made claims of soundproofing that has been designed into the buildings. Discussed below are the most common sources of intrusive noise leaks within buildings that can cause annoyance and lack of privacy. Practical solutions to the problems are also given.

A point to bear in mind when dealing with acoustic designs or specifications is that many materials and specifications are labeled in terms of "acoustic" designations (e.g., acoustic sealants, acoustic tiles, or acoustic doors). Labeling something with the "acoustic" qualification is completely meaningless without acoustic data relevant to its desired use (i.e., absorption coefficients and transmission loss [TL]) as backup. All things have acoustic properties. The key is to match the materials having the appropriate acoustic properties with the affected space to provide the desired results. Installation methods are also of critical importance to the effectiveness of the materials.

Noise from Heating, Ventilating and Air–Conditioning Systems

Heating, ventilating, and air conditioning (HVAC) systems can generate two types of noise problems within a building—they generate noise from their own mechanical equipment and they carry both that noise and other noise through their associated ductwork.

Mechanical Equipment

The noise-generating mechanical equipment associated with forced-air HVAC systems consists mostly of rotating fans. These fans generate pure tones and harmonics (tones at whole number multiples of the lowest

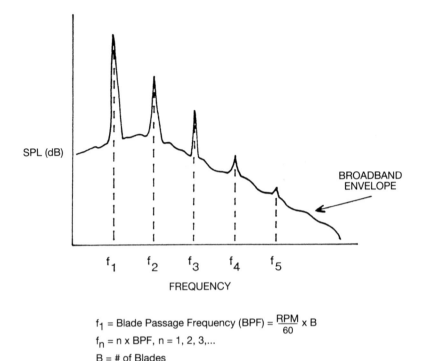

f_1 = Blade Passage Frequency (BPF) = $\dfrac{\text{RPM}}{60}$ x B

f_n = n x BPF, n = 1, 2, 3,...

B = # of Blades

FIGURE 7-1.   Fan noise spectrum. f_2, f_3, and so on, are harmonics of f_1, the fundamental blade passage frequency. SPL, Sound pressure level.

frequency pure tone) that are mathematically related directly to the number of rotating blades and their speed of rotation. A typical fan noise spectrum is illustrated in Fig. 7-1. If these tones are loud enough and of a high enough frequency to cause annoyance, the problem can be solved by lowering the rotational speed of the fans, which lowers the noise level and the frequency. Because frequencies below 1000 Hz are attenuated by the hearing mechanism by increasing amounts as the frequency decreases, the lower frequency tone would sound even quieter to people. Much research has been done on the effect of fan blade shape on noise generation. Thinning the blades and designing for optimal aerodynamic smoothness of airflow has proven to reduce noise levels slightly but practical limitations of performance dictate that noise will be generated as long as this type of design is utilized.

Noisy mechanical equipment should be completely enclosed in separate rooms from noise-sensitive areas. If ventilation is required for the equipment, it should be accomplished through nonsensitive areas, preferably to the outdoors.

Ductwork

There are many noise-generating phenomena associated with HVAC ductwork. In general, noise can travel through the inside of the duct or be radiated outside of the duct through the duct walls. The noise traveling inside the duct is usually generated by the rotating fans and the air flow through the duct. Standing waves can also be set up in straight ducts when the tones generated by the fans correspond to standing wave frequencies of the duct. The standing waves can be eliminated by adjusting the fan speed, treating the inside of the duct with acoustically absorptive material, or adding bends to the duct path. Tones higher than 1000 Hz can also be confined from reaching rooms far away from the noise-generating equipment by lining the ductwork with acoustically absorptive material and adding bends to the duct path.

Air flow noise is generally not characterized by pure tones and is thus classified as broadband in nature. Air flow noise is normally generated by turbulence. Turbulence in the air flow is usually generated by sharp changes in the airflow path, such as sharp bends, encounters with vanes that are not aerodynamically smooth in design, or sharp cross-sectional area changes in

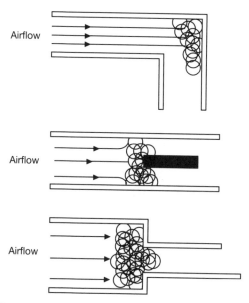

FIGURE 7-2.   Generation of turbulence by air flows in ducts.

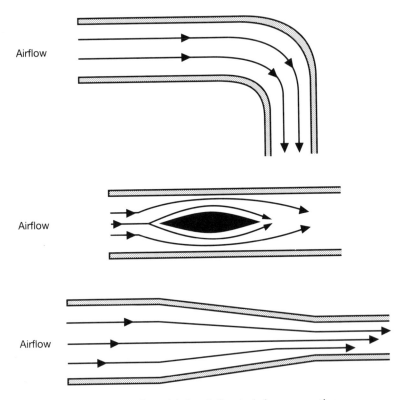

FIGURE 7-3.   Duct designs that minimize air flow turbulence generation.

the duct. This is illustrated in Fig. 7-2. Turbulence, and therefore noise generated by turbulence, can be minimized by ensuring smooth air flow conditions, using aerodynamic principles as shown in Fig. 7-3. This would also include aerodynamic designs of duct exit grills.

Noise radiated through the duct walls is known as breakout noise. Not only can mechanical equipment noise be carried through ductwork but conversations can also be carried through ductwork from one room to another. This can be the result of breakout noise or noise carried to an exit grill in another room. This can be minimized or eliminated by lining the duct interior with acoustically absorptive material or by wrapping the outside of the duct with lagging material, as was mentioned for noise control of industrial piping in Chapter 4. Lagging material is usually available in sheets and has high TL capabilities because of its design, using flexible (low stiffness) and heavy (high mass) materials. An effective lagging material is known as leaded vinyl. This is composed of a sandwich of a lead sheet surrounded by vinyl. Because lead has high mass per unit area and low stiffness, it makes an ideal material for high TL design as long as it is practical. Because of environmental considerations, lead has been replaced with other materials that have comparable TL ratings. Whenever duct lining, outside or inside the duct, is done it should completely cover the noise-emitting section to be most effective.

Plumbing Noise

Plumbing noise generation is similar to HVAC noise in that a fluid (in this case liquid; in the HVAC case, air) generates noise as it flows through a pipe (a duct in the HVAC case). Turbulence generated in the liquid flow by sharp bends in pipes, sharp changes in cross-sectional area, high fluid pressures, and pressure differentials at valves causes noise. As opposed to HVAC duct noise, the pipe interior cannot be treated with acoustically absorptive material to control the noise because such material would not survive the constant liquid exposure. Beside smoothing the flow and minimizing pressure differentials, the control of plumbing noise is limited to treatments external to the pipes.

Plumbing noise becomes an issue in multifamily unit dwellings because the pipes are often installed between walls and mounted vertically with little or no insulation between the pipes and the walls. In many cases, the pipes are even rigidly attached to the walls. The result is that, every time a toilet is flushed or water is used in a unit above another, the resultant water flow noise is heard clearly in the unit below. The most practical solution to this problem is to ensure that pipes are not rigidly attached to any walls and to wrap the pipes with lagging material. If pipes must be attached to walls for support, they can be attached through resilient materials or neoprene-type pads to break the vibration channel of the pipe noise to the wall.

Another point worth noting is that holes cut or drilled through floors for pipes are usually cut larger than necessary for the pipes to fit through. After the pipes are installed, leaving these holes unsealed provides an open path for sound travel. These openings may be the reason conversations can be heard clearly from bathrooms of units above or below each other. When the holes are sealed, the material that contacts the pipes directly should be neoprene or some resilient material to avoid structure-borne vibrations that could be carried from the pipes. Typical plumbing noise sources are illustrated in Fig. 7-4.

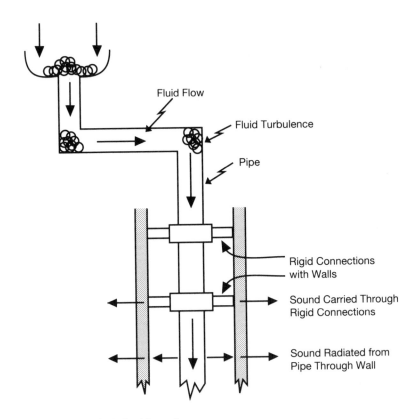

FIGURE 7-4.   Typical plumbing noise sources.

Privacy

The acoustic privacy provided by common building practices and materials is often limited because of the many potential noise leaks that are designed into most buildings. Because higher frequencies (above 1000 Hz) are effectively attenuated by most common building materials and lower frequencies (below 500 Hz) are not, speech intelligibility between units (requiring consonant recognition in the higher frequency range) would usually indicate a clear opening for sound passage between units. Included in this clear opening category would be HVAC ducts, radiators, or metal conduit carrying sound along their lengths. If only lower frequencies (bass sounds) are heard between units, there are probably no clear openings between units and the construction materials used do not provide adequate noise attenuation. If vibrations can be felt or things are buzzing or rattling because of noise gener-

TABLE 7-1   Common Treatments for Providing Acoustic Privacy between Dwellings

Symptom	Treatment
Conversation heard clearly between rooms	Completely seal open holes in walls and floors with nonhardening materials; break all solid vibration channels between rooms
Impact noise on hard floors from room above	Install carpet and padding on floor or float floor
Low frequencies (bass sounds) heard between rooms	Isolate vibrations with resilient materials, float floors, use multilayered partitions
Very loud source heard in adjacent room	Consider non-noise-sensitive buffer zone between loud room and room in question

ated in another unit, the problem is most likely structure-borne noise. Practical methods of correcting acoustic privacy problems are listed in Table 7-1.

Mechanical Equipment Noise

Mechanical equipment can generate airborne or structure-borne noise problems. As stated in Chapter 4, structure-borne noise can be solved principally by isolating the source from the building structure, using padding or tuned springs. If any mechanical equipment in a room having a common wall with a dwelling or occupied office unit generates more than 85 dBA at 3 ft, it should be enclosed or treated acoustically to ensure that it does not generate more than 85 dBA at 3 ft. Typical mechanical equipment noise sources and their control are listed in Table 7-2.

TABLE 7-2   Mechanical Equipment Noise Sources and Common Quieting Measures

Symptom	Treatment
Abnormal sounds such as squeals, scraping, flapping	Proper maintenance, balancing
Vibrations carried into building structure	Isolate equipment on tuned springs or pads, use active controls
Fan noise in ducted systems	Line ducts with absorptive material, provide smooth air flow path with bends, provide proper silencers, use active controls
High-level noise with normal operation	Enclose with materials that are effective in frequency range of interest, provide ventilation with lined ducts and louvers
Hissing noises	Seal leaks

Wind Noise

Whenever an opening or obstacle exists in an area of high-volume air flow, turbulence (and thus noise) is generated. The most common forms of wind noise are heard from pipes leaking steam, air under high pressure, or through windows cracked open on a windy day. In the case of the pipe, the high speed of the air flow through the small hole produces turbulence that results in the characteristic hissing noise. In the case of the window, turbulence is generated when a large volume of air from the outside tries to enter the building, at high speed, through the small opening of the partially open window. With the window only cracked open, the noise generated can have high-frequency components. With an opening that stays the same size, the frequency of wind noise usually increases with increases in wind speed. As the window is opened wider, the turbulence is lessened and the frequency lowers until the noise disappears completely when the air flow is relatively smooth into the room. You should not need this book to tell you that whistling windows can obviously be cured by sealing them to prevent airflow from coming through the window.

Some special cases of wind noise are worth mentioning. When a high-speed air flow passes around an obstacle, tones (known as Aeolian tones) are generated with dominant frequencies that depend on the size of the obstacle and the speed of the air flow. These are the tones that can be heard when wind blows through bare tree branches, screens, or hanging telephone cables. When air speeds are high enough, Aeolian tones can be generated through duct exit grills in HVAC systems. In these cases, air flow speeds must be kept to a minimum and exit grills must be designed to keep the air flow aerodynamically smooth.

Amplified Music

As with all other types of noise sources, low-frequency (bass) amplified music is the most difficult sound frequency range to control within buildings. As is also the case with the other sources, if the entire music noise spectrum can be heard between offices or dwelling units there is most likely an opening between the common walls that must be sealed. If only low-frequency noise is passing through, isolation of units is the only effective method to eliminate the noise transmission problem unless active vibration cancellation can be practically implemented at the source. Remember from Chapter 4 that the solution is not to add absorptive treatment (e.g., dropped ceilings or blankets). Using typical absorptive materials to control low-frequency noise transmission would provide a completely ineffective solution.

OFFICES

There are generally two types of offices: open plan and closed plan. The open-plan office is not completely enclosed and is either completely in the open, with no acoustic shielding from others, or surrounded by partial barriers for visual privacy from others. No significant noise attenuation or acoustic privacy can be expected from an open-plan office. If the partial barriers have absorptive materials on them, some reverberant field noise reduction may result by adding the barriers to the room but minimal acoustic privacy should be anticipated.

Closed-plan offices are completely enclosed and, because of that, are capable of providing sound privacy. However, because of economics and compromises in designs, the average closed-plan office does not provide much in the line of sound privacy. When sound privacy is an issue for a closed-plan office, several points must be kept in mind. The first is that all walls must be completely sealed at all perimeters, with no air gaps at wall edges, doors, windows, or any wall penetration. Doors should seal onto solid thresholds as opposed to carpet or other surfaces that can vary in thickness with use. The STC values of windows and doors must be comparable to that of the wall if maximum potential privacy is an issue.

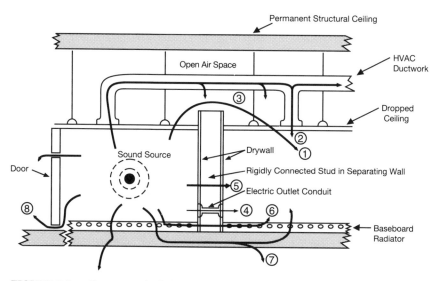

FIGURE 7-5. Common noise leaks occur (1) through dropped ceilings and around walls not extended to structural ceilings, (2) through duct grills, (3) through duct walls, (4) through conduit rigidly connected between rooms, (5) through studs rigidly connected between rooms, (6) through baseboards rigidly connected between rooms, (7) through floors or other building structures, and (8) around doors and windows.

TABLE 7-3   Practical Noise Control Measures for Offices

Problem	Treatment
Sound carried through dropped ceiling over separating wall into next room	Continue partition wall to structural ceiling or hang heavy curtain from structural ceiling to top of wall
Sound carried through ductwork	Line ducts with absorptive materials
Sound radiated from duct walls	Cover ducts with lagging materials
Sound traveling through conduit	Stagger electrical outlets for no solid connections or put resilient materials in conduit to break vibration chain
Sound traveling through studs between sides of partition wall	Stagger studs or install resilient pads between studs and wall connections
Sound traveling though baseboard radiator	Line with appropriate absorptive materials or isolate baseboard room sections
Structure-borne vibrations	Isolate room or float floor (if floor is a significant vibration channel)
Sound leaks around door openings	Seal doors with soft materials meeting solid materials around entire perimeter
Quiet office or large open-plan office, needs privacy without additional acoustic measures	Add sound-masking system

A common cost-cutting measure used in closed-plan offices is, when there is a dropped ceiling in the room, to bring the room walls up to the dropped ceiling and leave open air space between the dropped ceiling and the permanent ceiling. The typical dropped ceiling is composed of foam or porous board tiles that provide some acoustic absorption but minimal transmission loss. The office walls are then partial barriers because the sound travels over them and into the next room with a maximum of 15 to 25 dBA of attenuation (depending on the dropped ceiling material and installation), regardless of the wall material. This is another case of visual privacy not implying, and not providing, acoustic privacy. The most effective solution to this problem would be to continue the office walls above the dropped ceiling to seal with the permanent ceiling. An alternative solution that can result in a total wall attenuation of up to 45 dBA would be to hang leaded vinyl-type curtains from the permanent ceiling to the dropped ceiling meeting with the top of the office wall. The common practice of adding absorptive blankets above the dropped ceiling may provide an additional 5 to 10 dBA of attenuation but is limited by the weight support constraints of the dropped ceiling.

HVAC systems, solid conduits, and any other connections between rooms must be lined with absorptive material and isolated between offices when acoustic privacy is essential. Figure 7-5 illustrates common closed-office noise leaks and Table 7-3 lists ameliorative methods for these problems.

MULTIFAMILY DWELLINGS

Common noise sources to look for in multifamily dwellings are floor–ceiling penetrations and common wall penetrations. If these penetrations are not sealed well and the equipment penetrating the structures is not sufficiently isolated between units, significant sound leaks will result. If concrete slabs are not installed between floors, footsteps may be heard from units above, especially if carpets or padding have not been installed. Typical masonry block firewalls used between multifamily units provide STC values of at least 50 when combined with drywall on either side. This would provide adequate acoustic privacy. However, if typical wood joist and drywall designs are used between floors, as is the case in typical single family homes, minimal privacy will be provided.

SINGLE FAMILY DWELLINGS

A common complaint among single family homeowners with children is that they can hear the television, stereo, musical instruments, or telephone conversations of their children between floors. This is true for houses of any cost, especially for new houses built today, because no concern is given to acoustic privacy between rooms and floors unless people specifically ask and pay extra for it. If acoustic privacy is required in some areas of the new house to be constructed, make sure that it is addressed in the building plans. Solving acoustic problems after a house has been built is much more difficult than building insulation measures into the house as it is built. If privacy between floors is desired, consider isolating or floating the floor. If privacy between rooms is desired, consider the factors mentioned in previous sections of this chapter. A complete checklist of the most common sources of noise problems in multifamily housing units is provided below.

MULTIFAMILY DWELLING UNIT INSPECTION
CHECKLIST FOR AVOIDING NOISE PROBLEMS

Common/exterior walls
　Sealed around perimeters
　No holes
　Materials used that provide a minimum STC of 45 between units
　Pipes and ducts between walls insulated and isolated from walls
　Penetrations completely sealed with nonhardening materials

Common floors/ceilings

Sealed around perimeters

No holes

Materials used that provide a minimum IIC of 45 between units

Pipes and ducts between floor and ceiling insulated and isolated from floor and ceiling

Penetrations completely sealed with nonhardening materials

Exterior doors

Solid core

Sealed around perimeters with flexible materials

If in noisy area, door provides minimum STC of 40

If in noisy area, consider two doors with vestibule between them

Exterior windows

Window completely sealed to frame, frame completely sealed to wall with soft materials

If in noisy area, window provides minimum STC of 40 and year-round central air conditioning is available

If in noisy area, consider exterior windows in non-noise-sensitive rooms

General

Dwelling unit not near any mechanical equipment rooms, elevator shaft, or garbage chute

Building not near major noise sources such as highway, rail line, or airport flight path

Exterior wall of unit not facing commercial uses, industrial uses, or garbage pickup location

No nightclubs in same building or in sight of dwelling unit

If unit does not satisfy the above general categories, either ensure appropriate acoustic treatment or be aware that noise may be a problem

References

BIA. 1988. *Sound Insulation-Clay Masonry Walls*. TN-5A. Reston, VA: Brick Institute of America.

DuPree, R. B. 1988. *Catalog of STC and IIC Ratings for Wall and Floor/Ceiling Assemblies*. Berkeley, CA: Office of Noise Control, State of California Department of Health Services.

GA. 1992. *Fire Resistance Design Manual*. 13th ed. GA-600-92. Washington, D.C.: Gypsum Association.

NCMA. 1990. *Sound Transmission Class Ratings for Concrete Masonry Walls*. TEK 69B. Herndon, VA: National Concrete Masonry Association.

Appendix: Model Municipal Noise Ordinance

ARTICLE 1: PURPOSE

1.1. WHEREAS excessive sound is a serious hazard to the public health, welfare, safety, and the quality of life; and, WHEREAS a substantial body of science and technology exists by which excessive sound may be substantially abated; and, WHEREAS the people have a right to, and should be ensured an environment free from excessive sound, it is the policy of _____ to prevent excessive sound that may jeopardize the health, welfare, or safety of the citizens or degrade the quality of life.

1.2. This ordinance shall apply to the control of sound originating from stationary sources within the limits of _____

ARTICLE 2: DEFINITIONS

The following words and terms, when used in this ordinance, shall have the following meanings unless the context clearly indicates otherwise.

2.1. "Ambient Sound Level" is the total sound pressure level in the area of interest including the noise source of interest.

2.2. "A-Weighting" is the electronic filtering in sound level meters that models human hearing frequency sensitivity.

2.3. "Background Sound Level" is the total sound pressure level in the area of interest excluding the noise source of interest.

2.4.   "Commercial Area" is a group of commercial facilities and the abutting public right-of-way and public spaces.

2.5.   "Commercial Facility" is any premises, property, or facility involving traffic in goods or furnishing of services for sale or profit, including but not limited to:

   a. Banking and other financial institutions;
   b. Dining establishments;
   c. Establishments for providing retail or wholesale services;
   d. Establishments for recreation and entertainment;
   e. Office buildings;
   f. Transportation; and
   g. Warehouses.

2.6.   "Construction" is any site preparation, assembly, erection, repair, alteration or similar action, or demolition of buildings or structures.

2.7.   "C-Weighting" is the electronic filtering in sound level meters that models a flat response (output equals input) over the range of maximum human hearing frequency sensitivity.

2.8.   "dBA" is the A-weighted unit of sound pressure level.

2.9.   "dBC" is the C-weighted unit of sound pressure level.

2.10. "Decibel (dB)" is the unit of measurement for sound pressure level at a specified location.

2.11. "Emergency Work" is any work or action necessary to deliver essential services including, but not limited to, repairing water, gas, electric, telephone, sewer facilities, or public transportation facilities, removing fallen trees on public rights-of-way, or abating life-threatening conditions.

2.12. "Impulsive Sound" is a sound having a duration of less than 1 s with an abrupt onset and rapid decay.

2.13. "Industrial Facility" is any activity and its related premises, property, facilities, or equipment involving the fabrication, manufacture, or production of durable or nondurable goods.

2.14. "Motor Vehicle" is any vehicle that is propelled or drawn on land by an engine or motor.

2.15. "Muffler" is a sound-dissipative device or system for attenuating the sound of escaping gases of an internal combustion engine.

2.16. "Multidwelling Unit Building" is any building wherein there are two or more dwelling units.

2.17. "The Municipality" is (name of municipality in question).

2.18. "Noise" is any sound of such level and duration as to be or tend to be injurious to human health or welfare, or which would unreasonably interfere with the enjoyment of life or property throughout the Municipality or in any portions thereof, but excludes all aspects of the employer–employee relationship concerning health and safety hazards within the confines of a place of employment.

2.19. "Noise Control Administrator (NCA)" is the noise control officer designated as the official liaison with all municipal departments, empowered to grant permits for variances.

2.20. "Noise Control Officer (NCO)" is an officially designated employee of the Municipality trained in the measurement of sound and empowered to issue a summons for violations of this ordinance.

2.21. "Noise Disturbance" is any sound that (a) endangers the safety or health of any person, (b) disturbs a reasonable person of normal sensitivities, or (c) endangers personal or real property.

2.22. "Person" is any individual, corporation, company, association, society, firm partnership, joint stock company, the Municipality or any political subdivision, agency or instrumentality of the Municipality.

2.23. "Public Right-of-Way" is any street, avenue, boulevard, road, highway, sidewalk, or alley that is leased, owned, or controlled by a governmental entity.

2.24. "Public Space" is any real property or structures thereon that is owned, leased, or controlled by a governmental entity.

2.25. "Pure Tone" is any sound that can be judged as a single pitch or set of single pitches by the NCO.

2.26. "Real Property Line" is either (a) the imaginary line, including its vertical extension, that separates one parcel of real property from another, or (b) the vertical and horizontal boundaries of a dwelling unit that is one in a multidwelling unit building.

2.27. "Residential Area" is a group of residential properties and the abutting public rights-of-way and public spaces.

2.28. "Residential Property" is property used for human habitation, including but not limited to:
   a. Private property used for human habitation;
   b. Commercial living accommodations and commercial property used for human habitation;
   c. Recreational and entertainment property used for human habitation; and
   d. Community service property used for human habitation.

2.29. "Sound Level" is the instantaneous sound pressure level measured in decibels with a sound level meter set for A-weighting on slow integration speed, unless otherwise noted.

2.30. "Sound Level Meter (SLM)" is an instrument used to measure sound pressure levels conforming to Type 1 or Type 2 standards as specified in ANSI Standard S1.4-1983 or the latest version thereof.

2.31. "Sound Pressure Level (SPL)" is 20 multiplied by the logarithm, to the base 10, of the measured sound pressure divided by the sound pressure associated with the threshold of human hearing, in units of decibels.

2.32. "Weekday" is any day, Monday through Friday, that is not a legal holiday.

ARTICLE 3: POWERS, DUTIES, AND QUALIFICATIONS OF THE NOISE CONTROL OFFICERS AND ADMINISTRATORS

3.1. The provisions of this ordinance shall be enforced by the noise control officers (NCOs).

3.2. The noise control administrator (NCA) shall have the power to:
   a. Coordinate the noise control activities of all municipal departments and cooperate with all other public bodies and agencies to the extent practicable;
   b. Review the actions of other municipal departments and advise such departments of the effect, if any, of such actions on noise control;
   c. Review public and private projects, subject to mandatory review or approval by other departments or boards, for compliance with this ordinance; and
   d. Grant permits for variances according to the provisions of Article 9.

3.3. A person shall be qualified to be an NCO if the person has satisfactorily completed any of the following:
   a. An instructional program in community noise from a certified noise control engineer, as evidenced by certification from the Institute of Noise Control Engineering (INCE);
   b. An instructional program in community noise from another NCO; or
   c. Education or experience or a combination thereof certified by the NCA as equivalent to the provisions of (a) or (b) of this section.

3.4. Noise measurements taken by an NCO shall be taken in accordance with the procedures specified in Article 5.

ARTICLE 4: DUTIES AND RESPONSIBILITIES OF OTHER DEPARTMENTS

4.1. All departments and agencies of the Municipality shall carry out their programs according to law and shall cooperate with the NCA in the implementation and enforcement of this ordinance.

4.2. All departments charged with new projects or changes to existing projects that may result in the production of noise shall consult with the NCA prior to the approval of such projects to ensure that such activities comply with the provisions of this ordinance.

ARTICLE 5: SOUND MEASUREMENT PROCEDURES

5.1. Insofar as practicable, sound will be measured while the source under investigation is operating at normal, routine conditions and, as necessary, at other conditions, including but not limited to, design, maximum, and fluctuating rates.

5.2. All tests shall be conducted in accordance with the following procedures:

    a. The NCO shall, to the extent practicable, identify all sources contributing sound to the point of measurement.

    b. Measurements shall be taken at or within the property line of the affected person.

    c. The SLM must be calibrated using a calibrator recommended by the SLM manufacturer before and after each series of readings and at least once each hour.

    d. The SLM must be recertified and the calibrator must be recalibrated at least once each year by the manufacturer or by a person that has been approved by the NCA. A copy of written documentation of such recertification and recalibration, in a form approved by the NCA, shall be kept with the equipment to which it refers.

    e. No outdoor measurements shall be taken:

        1. During periods when wind speeds (including gusts) exceed 15 mph;

        2. Without a windscreen, recommended by the SLM manufacturer, properly attached to the SLM;

        3. Under any condition that allows the SLM to become wet (e.g., rain, snow, or condensation); or

        4. When the ambient temperature is out of the range of the tolerance of the SLM.

5.3. The report for each measurement session shall include:

    a. The date, day of the week, and times at which measurements are taken;

    b. The times of calibration;

    c. The weather conditions;

    d. The identification of all monitoring equipment by manufacturer, model number, and serial number;

e. The normal operating cycle of the sources in question with a description of the sources;

f. The ambient sound level, in dBA, with the sources in question operating;

g. The background sound level, in dBA, without the sources in question operating; and

h. A sketch of the measurement site, including measurement locations and relevant distances, containing sufficient information for another investigator to repeat the measurements under similar conditions.

5.4. Prior to taking noise measurements the investigator shall explore the vicinity of the source in question to identify any other sound sources that could affect measurements, to establish the approximate location and character of the principal sound source, and to select suitable locations from which to measure the sound from the source in question.

5.5. When measuring continuous sound, or sound that is sustained for more than 1 s at a time, the SLM shall be set for A-weighting, slow meter response speed, and the range (if the SLM is designed to read levels over different ranges of SPLs) shall be set to that range in which the meter reads closest to the maximum end of the scale. When the measured sound level is variable or fluctuating over a range greater than ± 3 dBA, using the slow meter response speed, the fast meter response speed shall be used. In either case, both the minimum and maximum readings shall be recorded to indicate the range of monitored values.

5.6. The SLM shall be placed at a minimum height of 3 ft above the ground or from any reflective surface. When handheld, the microphone shall be held at arm's length and pointed at the source at the angle recommended by the SLM manufacturer.

5.7. If extraneous sound sources, such as aircraft flyovers or barking dogs, that are unrelated to the measurements increase the monitored sound levels, the measurements should be postponed until these extraneous sounds have become of such a level as not to increase the monitored sound levels of interest.

5.8. The monitoring session should last for a period of time sufficient to ensure that the sound levels measured are typical of the source in question, but in no event shall the duration of testing be less than 5 min.

5.9. The background sound levels shall be subtracted from the measured sound levels of the source of interest by using Table 1 to determine the sound levels from the source of interest alone. If the ambient sound level is less than 3 dBA higher than the background sound level, the source level cannot be derived and a violation of the ordinance cannot be substantiated.

TABLE 1   Correction for Background Levels[a]

Difference between Ambient and Background Sound Levels	Correction Factor to Be Subtracted from Ambient Level for Source Level
3	3
4, 5	2
6–9	1
10 or more	0

[a] In dBA.

ARTICLE 6: SOUND LEVEL LIMITATIONS

6.1. No person shall cause, suffer, allow, or permit the operation of any sound source on a particular category of property or any public space or right-of-way in such a manner as to create a sound level that exceeds the background sound level by at least 10 dBA during daytime (7:00 A.M. to 10:00 P.M.) hours and by at least 5 dBA during nighttime (10:00 P.M. to 7:00 P.M.) hours when measured at or within the real property line of the receiving property, except as provided in Section 6.1.1. Such a sound source would constitute a noise disturbance.

    6.1.1. If the background sound level cannot be determined, the absolute sound level limits set forth in Table 2 shall be used.

    6.1.2. If the sound source in question is a pure tone, the limits of Table 2 shall be reduced by 5 dBA.

TABLE 2   Maximum Permissible Sound Levels[a]

	Receiving Property		
	Residential		Commercial
Source Property	7:00 A.M.–10:00 P.M.	10:00 P.M.–7:00 A.M.	(All Times)
Residential	55	50	65
Commercial	65	50	65
Industrial	65	50	65

[a] In dBA. These levels would be appropriate for typical suburban environments. Urban environments may allow for limits that are 5 to 10 dBA higher and rural or quiet suburban environments may allow for limits that are 5 to 10 dBA lower than those listed. The specific limitations should be based on the environment and tastes of the municipality.

6.1.3. Nonrepetitive impulsive sound sources shall not exceed 90 dBA or 120 dBC at or within a residential real property line, using the fast meter response speed.

6.1.4. In multidwelling unit buildings, if the background sound level cannot be determined, the daytime limit is 45 dBA and the nighttime limit is 35 dBA for sounds originating in another dwelling within the same building.

6.2. The following are exempt from the sound level limits of Section 6.1:

a. Noise from emergency signaling devices;

b. Noise from an exterior burglar alarm of any building provided such burglar alarm shall terminate its operation within 5 min of its activation;

c. Noise from domestic power tools, lawn mowers, and agricultural equipment when operated between 8:00 A.M. and 8:00 P.M. on weekdays and between 9:00 A.M. and 8:00 P.M. on weekends and legal holidays, provided they generate less than 85 dBA at or within any real property line of a residential property;

d. Sound from church bells and chimes when a part of a religious observance or service;

e. Noise from construction activity provided all motorized equipment used in such activity is equipped with functioning mufflers, except as provided in Section 7.2(f);

f. Noise from snow blowers, snow throwers, and snow plows when operated with a muffler for the purpose of snow removal.

ARTICLE 7: SPECIFIC PROHIBITED ACTS

7.1. No person shall cause, suffer, allow, or permit to be made verbally or mechanically any noise disturbance, as defined in Section 6.1.

7.2. No person shall cause, suffer, allow, or permit the following acts:

a. Operating, playing, or permitting the operation or playing of any radio, television, phonograph, or similar device that reproduces or amplifies sound in such a manner as to create a noise disturbance (as defined in Section 6.1) for any person other than the operator of the device;

b. Using or operating any loudspeaker, public address system, or similar device between 10:00 P.M. and 8:00 A.M. the following day, such that the sound therefrom creates a noise disturbance (as defined in Section 6.1) across a residential real property line;

c. Owning, possessing, or harboring any animal or bird that, frequently or for continued duration, generates sounds that create a noise disturbance (as defined in Section 6.1) across a residential real property line;

d. Loading, unloading, opening, closing, or other handling of boxes, crates, containers, building materials, liquids, garbage cans, refuse, or similar objects, or the pneumatic or pumped loading or unloading of bulk materials in liquid, gaseous, powder, or pellet form, or the compacting of refuse by persons engaged in the business of scavenging or garbage collection, whether private or municipal, between 9:00 P.M. and 7:00 A.M. the following day on a weekday and between 9:00 P.M. and 9:00 A.M. the following day on a weekend day or legal holiday except by permit, when the sound therefrom creates a noise disturbance (as defined in Section 6.1) across a residential property line;

e. Operating or permitting the operation of any motor vehicle whose manufacturer's gross weight rating is in excess of 10,000 lb, or any auxiliary equipment attached to such a vehicle, for a period of longer than 5 min in any hour while the vehicle is stationary, for reasons other than traffic congestion or emergency work, on a public right-of-way or public space within 150 ft of a residential area between 8:00 P.M. and 8:00 A.M. the following day;

f. Operating or permitting the operation of any tools or equipment used in construction, drilling, earthmoving, excavating, or demolition work between 6:00 P.M. and 7:00 A.M. the following day on a weekday or at any time on a weekend day or legal holiday, except for emergency work, by variance issued pursuant to Article 9, or when the sound level does not exceed any applicable relative or absolute limit specified in Section 6.1.

ARTICLE 8: EXEMPTIONS

8.1. The provisions of this ordinance shall not apply to:

a. The generation of sound for the purpose of alerting persons to the existence of an emergency except as provided in Section 6.2(b);

b. The generation of sound in the performance of emergency work; or

c. The generation of sound in situations within the jurisdiction of the Federal Occupational Safety and Health Administration.

8.2. Noise generated from municipally sponsored or approved celebrations or events shall be exempt from the provisions of this ordinance.

ARTICLE 9: VARIANCE CONDITIONS

9.1. Any person who owns or operates any stationary noise source may apply to the NCA for a variance from one or more of the provisions of this ordinance. Applications for a permit of variance shall supply information including, but not limited to:

    a. The nature and location of the noise source for which such application is made;

    b. The reason for which the permit of variance is requested, including the hardship that will result to the applicant, his/her client, or the public if the permit of variance is not granted;

    c. The level of noise that will occur during the period of the variance;

    d. The section or sections of this ordinance for which the permit of variance shall apply;

    e. A description of interim noise control measures to be taken for the applicant to minimize noise and the impacts occurring therefrom; and

    f. A specific schedule of the noise control measures that shall be taken to bring the source into compliance with this ordinance within a reasonable time.

    9.1.1. Failure to supply the information required by the NCA shall be cause for rejection of the application.

    9.1.2. A copy of the permit of variance must be kept on file by the municipal clerk for public inspection.

9.2. The NCA may charge the applicant a fee of $_____ to cover expenses resulting from the processing of the permit of variance application.

9.3. The NCA may, at his/her discretion, limit the duration of the permit of variance, which shall be no longer than 1 year. Any person holding a permit of variance and requesting an extension of time shall apply for a new permit of variance under the provisions of this section.

9.4. No variance shall be approved unless the applicant presents adequate proof that:

    a. Noise levels occurring during the period of the variance will not constitute a danger to public health; and

    b. Compliance with the ordinance would impose an unreasonable hardship on the applicant without equal or greater benefits to the public.

9.5. In making the determination of granting a variance, the NCA shall consider:

    a. The character and degree of injury to, or interference with, the health and welfare or the reasonable use of property that is caused or threatened to be caused;

    b. The social and economic value of the activity for which the variance is sought; and

    c. The ability of the applicant to apply the best practical noise control measures.

9.6. The permit of variance may be revoked by the NCA if the terms of the permit of variance are violated.

9.7. A variance may be revoked by the NCA if there is:

    a. Violation of one or more conditions of the variance;

    b. Material misrepresentation of fact in the variance application; or

    c. Material change in any of the circumstances relied on by the NCA in granting the variance.

ARTICLE 10: ENFORCEMENT PROCEDURES

10.1. Violation of any provision of this ordinance shall be cause for a summons to be issued by the NCO according to procedures set forth in (*Administrative Code reference*).

10.2. In lieu of issuing a summons as provided in Section 10.1, the NCO may issue an order requiring abatement of any sound source alleged to be in violation of this ordinance within a reasonable time period and according to guidelines that the NCO may prescribe.

10.3. Any person who violates any provision of this ordinance shall be subject to a fine for each offense of not more than $_____.

    10.3.1. If the violation is of a continuing nature, each day during which it occurs shall constitute an additional, separate, and distinct offense.

10.4. No provision of this ordinance shall be construed to impair any common law or statutory cause of action, or legal remedy therefrom, of any person for injury or damage arising from any violation of this ordinance or from other law.

ARTICLE 11: SEVERABILITY

11.1. If any provision of this ordinance is held to be unconstitutional, preempted by federal law, or otherwise invalid by any court of competent jurisdiction, the remaining provisions of the ordinance shall not be invalidated.

ARTICLE 12: EFFECTIVE DATE

12.1. This ordinance shall take effect on _____.

Sources
Adapted from the following references:
Anon. 1977. *Model Noise Control Ordinance for Stationary Sources.* Trenton, NJ: New Jersey State Department of Environmental Protection and Energy.
Anon. 1977. *Model Community Noise Control Ordinance.* Berkeley, CA: California Department of Health Services Office of Noise Control.

Glossary

Absorption Coefficient (α)   the dimensionless ratio of absorbed to incident sound energy from a single interaction between a sound wave and a partition. Values range from 0 to 1.

Absorption (Sound)   the product of absorption coefficient and surface area of a material, in units of sabins, used to designate the amount of sound absorbed by that material.

Acoustics   the science or study of sound.

Acoustic Trauma   the physical destruction of the inner ear hearing organs resulting from exposure to peak sound pressure levels greater than 140 dB. Acoustic trauma normally results in permanent hearing loss.

Active Noise Control   the cancellation of sound waves by introducing a mirror image (180° out of phase) of the original sound wave to the sound path.

Ambient Noise Level   the total noise level in the acoustic environment, usually including the noise source of interest.

American National Standards Institute (ANSI)   a voluntary federation of U.S. organizations concerned with the development of standards. ANSI standards are drafted by committees of industry experts and published only after board review and determination of national consensus.

American Society for Testing and Materials (ASTM)   a voluntary federation of U.S. organizations concerned with the development of standard testing methods. ASTM standards are drafted by committees of industry experts and published only after determination of national consensus.

Anechoic Chamber   a room having terminations (walls, floor, and ceiling) that absorb all sound incident on them; used in laboratories to measure direct sound fields from sources.

Attenuation   reduction in level.

A-Weighting   electronic filtering in sound level meters that models human hearing frequency sensitivity.

Background Noise Level   the noise level in the acoustic environment, usually excluding the noise source of interest.

Bandwidth (BW)   the frequency range of maximum flat filter response in an instrument, the upper and lower frequency limits of which occur where the response drops off by 3 dB.

Broadband Spectrum   an SPL vs. frequency plot having no discrete frequency dominance or peaks, varying smoothly with frequency.

Calibration   using an instrument emitting an accepted SPL and frequency as a reference for ensuring that the monitoring instruments provide reliable results.

Coincidence Frequency   the bending wave resonance frequency of a partition, dependent on the material and thickness, that causes a reduction in TL effectiveness in a narrow frequency range around it.

Community Noise Equivalent Level (CNEL)   a 24-h continuous L_{eq} with 5 dBA added to levels occurring between 7:00 P.M. and 10:00 P.M. and 10 dBA added to levels occurring between 10:00 P.M. and 7:00 A.M.. The added values are used to account for added sensitivity during evening and typical nighttime sleeping hours.

Continuous Sound   sound having a steady, nonimpulsive nature.

Coverage Angle (C_ζ)   the angle, with respect to the front side of a loudspeaker, between locations at which the emitted sound level sensitivity drops off by 6 dB.

Critical Frequency   the frequency at which coincidence frequency effects begin to be noticed for a partition.

C- Weighting   electronic filtering in sound level meters that models a flat response (output equals input) over the range of maximum human hearing sensitivity.

Day-Evening-Night Sound Level (L_{den})   the same as CNEL.

Day-Night Sound Level (L_{dn} or DNL)   a 24-hr continuous L_{eq} with 10 dBA added to levels occurring between 10:00 P.M. and 7:00 A.M. to account for greater sensitivity during typical sleeping hours.

dBA   A-weighted unit of sound pressure level.

dBC   C-weighted unit of sound pressure level.

Decibel (dB)   a unit of sound level implying 10 multiplied by a logarithmic ratio of power or some quantity proportional to power. The logarithm is to the base 10.

Diffraction   the act of sound waves traveling around barriers, especially pronounced when the sound wavelength size is comparable to or greater than the dimensions of the barriers.

Diffuse Field   area, within a room, where SPLs do not vary significantly with location. Diffuse fields are caused by reverberation.

Diffusion   the act of sound waves spreading out over a wide area after reflecting off of a convex or uneven surface.

Direct Field   area where the sound measured can be attributed to the source alone without effects of reflections off of walls or obstructions.

Directivity   the spherical coverage angle characteristics of a source.

Directivity Factor (Q)   the ratio of the intensity of a sound source near large reflective surfaces to the intensity when radiating into open space.

Directivity Index (D) $10 \times \log(Q)$, the amount added to the SPL of a spherically radiating source when placed near large reflective surfaces causing the directivity factor used.

Echo   the perception of two distinct sounds resulting from the difference in arrival times of sound waves traveling over different paths but originating from a single source.

Effective Perceived Noise Level (EPNL)   a rating method used by the FAA for certifying aircraft in terms of the annoyance of aircraft flyovers (in units of EPNdB).

Equivalent Sound Level (L_{eq})   an SPL that, if constant over a specified time period, would contain the same sound energy as the actual sound that varies in level with time. The reference time period is usually specified in terms of hours in parentheses (e.g., $L_{eq(1)}$ refers to a 1-h L_{eq} value).

Exchange Rate   the dBA level associated with a change of noise exposure duration by a factor of 2.

Far Field   area outside of the near field, where measurements can be reproduced with consistency.

Field Impact Insulation Class (FIIC)   an IIC rating using values measured in actual installations rather than in a laboratory.

Field Sound Transmission Class (FSTC)   an STC rating using values measured in actual installations rather than in a laboratory.

Free Field   area having no obstructions or reflective surfaces in the sound propagation path.

Frequency (f)   the rate, in hertz (cycles per second), at which periodic (sinusoidal) acoustic pressure oscillations occur. Frequency is interpreted subjectively as pitch. Humans can hear sounds having frequencies between 20 and 20,000 Hz.

Fresnel Number (N)   used in partial barrier analysis, a value defined in acoustic diffraction theory, and based on the locations of a source, receiver, and the top of a barrier.

Fundamental Frequency   the lowest resonance frequency of a system.

Harmonic   a positive integer multiple of the fundamental acoustic resonance frequency, including the fundamental. The first harmonic corresponds to the fundamental and the second harmonic corresponds to the first overtone.

Hearing Protection Device (HPD)   a device, typically either a muff or plug, that covers the ear canal to reduce noise levels before the sound enters the hearing mechanism.

Helmholtz Resonator   also known as a volume resonator, a device having a small opening leading to a larger volume of air used to amplify or absorb sound frequencies covering a limited range.

Impact Insulation Class (IIC)   a single number rating system for the sound attenuation effectiveness of floor–ceiling assemblies on impact noises, in which SPLs measured from using a tapping machine are matched to a standard curve. IIC measurements are performed in laboratory-type environments.

Impedance (Acoustic)   the ability of a medium to restrict the flow of acoustic energy, related to the cross-sectional area of the propagation path. When the acoustic impedance of a new medium is the same as that of the first medium, acoustic energy flows through unabated; when there is a change in impedance between media, there is an impedance mismatch and energy flow is restricted.

Impulsive Sounds   sounds that last less than 1 s having a sudden start and end.

Infrasound   sound waves having dominant frequency components below 20 Hz, the lower frequency limit of human hearing sensitivity.

Insertion Loss (IL)   the difference, in decibels, between the SPL before and after a sound-attenuating device is placed in the path between the source and receiver.

Inverse Square Law   the condition in open spherical wave sound propagation from a point source in which intensity drops off as the reciprocal of the square of the distance from the source. This translates to the ideal condition that SPL drops off at a rate of 6 dB per doubling of distance from the source.

Line Source   a sound source composed of many point sources in a defined line, such as a steady stream of traffic on a highway or a long train.

Mass Law   a relationship that relates a doubling in mass or frequency to a 6-dB increase in TL for a homogeneous partition over a specific frequency range.

Narrow Band Analyzer   a spectrum analyzer that measures sound levels in terms of frequency bands smaller than octave bands.

Near Field   area, close to a sound source, where sound measurements fluctuate dramatically.

Noise   unwanted sound.

Noise Abatement Criteria (NAC)   noise level limits, in terms of $L_{eq(1)}$ or $L_{10(1)}$, promulgated by Federal Highway Administration regulations for vehicular traffic noise generated by the construction of new highways or the expansion of existing ones.

Noise Isolation Class (NIC)   a single number rating of the sound attenuation effectiveness of a partition based on matching NR_{TL} values to a standard curve.

Noise Level Reduction (NLR)   the outdoor-to-indoor attenuation of noise levels afforded by the exterior wall of a building. The NLR is used only in FAA mitigation recommendations.

Noise Reduction (NR)   the reduction of sound level within a room caused by adding absorptive material to the room.

Noise Reduction (NR_{TL})   a measure of the sound attenuation effectiveness of a partition, the difference between average SPLs in two rooms with a noise source of interest in one of the rooms.

Noise Reduction Coefficient (NRC)   a single number rating system for absorption coefficients over the speech frequency range. The NRC is defined mathematically as the arithmetic average of the absorption coefficients at 250, 500, 1000, and 2000 Hz.

Noise Reduction Rating (NRR)   a rating, in decibels, of the effectiveness of HPDs on reducing noise levels reaching the hearing mechanism.

Noise-Sensitive Location   a defined area where human activity may be adversely affected when noise levels exceed predefined thresholds of acceptability or when levels increase by predefined thresholds of change.

Octave Band   a frequency band whose upper limit is twice the lower limit, and is identified by a geometric mean frequency, called the center frequency. Standard octave band center frequencies are defined in ANSI Standard S1.6-1984.

Octave Band Analyzer   an instrument that measures sound levels in terms of octave bands.

Omnidirectional Source   a source that emits equal amounts of energy in all directions, generating spherical waves.

Outdoor–Indoor Transmission Class (OITC)   an A-weighted rating of the sound reduction effectiveness of a partition that separates an indoor from an outdoor environment.

Overtone   a positive integer multiple of a fundamental acoustic resonance frequency, beginning at twice the fundamental.

Passenger Car Equivalent (PCE)   a unit used in traffic studies to account for the fact that trucks emit higher noise levels than cars.

Percent Highly Annoyed (%HA)   parameter used to rate annoyance against L_{dn} noise levels.

Percentile Levels (L_n, $0 < n < 100$)   the percentage of observation time that a certain SPL has been exceeded. For example, L_{10} corresponds to the SPL exceeded 10% of the observation time. The observation time is usually specified in terms of hours in parentheses (e.g., $L_{10(1)}$ refers to a 1-h L_{10} value).

Point Source   a source whose dimensions are small compared to propagation distances described in reference to it.

Presbycusis   hearing loss attributed to the aging process.

Pure Tone   a sound dominated by energy in a single frequency.

Reflection   the act of sound bouncing off of a partition, usually occurring from smooth, hard surfaces.

Refraction   the act of sound waves bending or changing propagation direction as they travel from one medium or medium condition (such as temperature, density, humidity, or wind current) into another.

Resonance (Acoustic or Room)   the generation of standing waves within a space at specific frequencies that correlate certain fractions of wavelengths, and integer multiples of them, with the dimensions of the space.

Resonance (Mechanical)   an increase in response of a material at a specific frequency that is dependent on the physical characteristics of the material.

Reverberant Field   same as diffuse field.

Reverberation   the amplification of sound within an enclosed space caused by multiple reflections off reflective terminations (i.e., walls, ceilings, floors, or obstacles) of the room.

Reverberation Chamber   a room having terminations (walls, floor, and ceiling) that reflect all sound incident on them; used in laboratories to set up diffuse sound fields from sources.

Reverberation Time (RT_{60} or T_{60})   the time, in seconds, it takes for the SPL in a room to decrease by 60 dB after a sound source (emitting levels more than 60 dB above the background level) has stopped emitting sound.

Shadow Zone   an area below which sound waves have bent upward because of atmospheric conditions. In such areas a distant sound source would not be as loud as expected.

Sociocusis   hearing loss attributed to nonoccupational, environmental factors.

Sound Concentration   the focusing of sound waves caused by reflections from concave surfaces or any other surface that causes sound waves to focus at a particular location.

Sound Exposure Level (SEL)   a rating, in decibels, of a discrete event, such as an aircraft flyover or train passby, that compresses the total sound energy of the event into a 1-s time period.

Sound Intensity   a quantity that describes sound in terms of both magnitude and direction of propagation.

Sound Level Meter (SLM)   an instrument used to measure sound pressure levels.

Sound Power Level (L_W) $10 \times \log(W/W_{ref})$, where $W =$ power and $W_{ref} = 1 \times 10^{-12}$ W.

Sound Pressure Level (SPL or L_p) $20 \times \log(p/p_{ref})$, where $p =$ root mean square acoustic pressure and $p_{ref} = 2 \times 10^{-5}$ N/m^2. p_{ref} corresponds to the pressure at the threshold of hearing.

Sound Transmission Class (STC)   a single number rating for a TL spectrum of a partition matched to a standard curve. STC measurements are performed in laboratory-type environments.

Spectrum   a graphical representation of sound level vs. frequency.

Spectrum Analyzer   a device that measures and manipulates spectra, available in many bandwidth possibilities. Octave band analyzers are the most common types of spectrum analyzers.

Speech Interference Level (SIL)   a single number rating for speech intelligibility, the arithmetic average of SPLs in the 500-, 1000-, 2000-, and 4000-Hz octave bands.

Standard Threshold Shift (STS)   an average loss of 10 dB in hearing ability at 2000, 3000, and 4000 Hz, on a temporary basis, caused by high-level noise exposure; also known as temporary threshold shift (TTS).

Threshold of Hearing (0 dBA)   the SPL below which sound cannot be heard by the average person with a healthy hearing mechanism.

Threshold of Pain (120 dBA)   the SPL over which sound causes physical pain to the average listener's ears.

Time-Weighted Average (TWA)   a noise exposure rating, in dBA, based on an 8-h L_{eq} with a 5-dBA exchange rate for OSHA compliance.

Tinnitus   a ringing or buzzing sound heard by an individual when the source of that sound is a medical abnormality or overextension in the auditory system.

Transducer   a device that transforms energy from one form into another.

Transmissibility   used in vibration analysis, the ratio of output to input energy transmitted from a vibrating source to another material.

Transmission Coefficient (τ)   the dimensionless ratio of transmitted to incident sound energy from a single interaction between a sound wave and a partition. Values range from 0 to 1.

Transmission Loss (TL)   a measure of the sound attenuation effectiveness of a partition, in units of decibels.

Ultrasound   sound waves having dominant frequency components above 20,000 Hz, the upper frequency limit of human hearing sensitivity.

Wavelength (w)   the distance between successive repeating portions of a pure tone sound wave.

Index